P9-CCZ-544

Earth Algebra

College Algebra with Applications to Environmental Issues

Earth Algebra

College Algebra with Applications to Environmental Issues

CHRISTOPHER SCHAUFELE
Kennesaw State College

NANCY ZUMOFF
Kennesaw State College

HarperCollins*CollegePublishers*

Sponsoring Editor: Anne Kelly
Developmental Editor: Kathy Richmond
Project Editor: Cathy Wacaser
Design Administrator: Jess Schaal
Text and Cover Design: Lesiak/Crampton Design Inc: Lucy Lesiak
Cover Photo: © 1990 Kevin Schafer/Martha Hill
Picture Research: Kelly Mountain/Roberta Knauf
Production Administrator: Randee Wire
Compositor: Interactive Composition Corporation
Text Printer and Binder: R.R. Donnelley & Sons Company
Cover Printer: R.R. Donnelley & Sons Company

♲ Printed on recycled paper.

Library of Congress Cataloging-in-Publication Data

Schaufele, Christopher.
Earth algebra: college algebra with applications to environmental issues/Christopher Schaufele, Nancy Zumoff.
p. cm.
Includes bibliographical references and index.
ISBN 0-06-500886-3
1. Carbon dioxide—Environmental aspects—Mathematical models. 2. Global warming—Mathematical models. 3. Environmental sciences—Mathematics. 4. Algebra—Study and teaching. I. Zumoff, Nancy. II. Title
TD885.5.O3S32 1995
363.7373′87—dc20 94-30342
CIP

95 96 97 9 8 7 6 5 4 3 2

Contents

Preface

APPROACH

This college algebra text focuses on modeling (primarily by curve fitting) real data concerning environmental issues, decision making, reading, writing, and oral reporting. Students work in small groups with graphing calculators or appropriate technology. After mathematical concepts and skills are reviewed, much of the work is done in student groups with the instructor serving more as a guide than a lecturer. Students write and present oral reports that summarize the work completed for each text section. These components—group work, student reports, and use of graphing calculators—make for a course in which students actively participate.

The environmental focus provides a mass of quantifiable data that is readily available and is of interest to both students and faculty. The mathematics used grows out of the need to answer particular questions, some of which have many possible answers. Students derive models for given data and use it to predict pertinent events. In a final project, students submit proposals to improve environmental conditions. With this approach students actually use equations to make decisions about real situations.

THE DEVELOPMENTAL STORY OF THE TEXT

We wrote the preliminary version of *Earth Algebra* in response to our departmental concern over problems encountered in the traditional college algebra course. We pondered the question of what makes college algebra boring to so many. One colleague suggested, "If you want to make a course interesting, then you should study something of interest." To do this we chose to couch most of the standard topics of college algebra in the vitally important issues of the environment. The course has been part of the Kennesaw State College curriculum since Fall 1991, and student reaction has been quite favorable. With the help of extensive reviews by 35 mathematics professors, we revised to produce a one-color preliminary text. This "first" edition is based on input from users of the preliminary version.

CHANGES FROM THE PREVIOUS EDITION

For the new edition we incorporate many ideas based on "user diaries" where people recorded their experiences using the book. For instance, based on user feedback, note the following:

- More discussion on the nature of modeling in the text before the first linear model in Chapter 3 and throughout the text; and more steps and guidance for students in the subsequent models.
- There is a modification of assumptions relating increased carbon dioxide concentration, temperature change, and ocean level that better reflects current evidence.
  - A brief discussion of units is included where appropriate.
- Chapters 10 and 16 of the preliminary version are combined for thorough coverage of logarithms and exponents.
- Elementary rules for solving inequalities are added to Chapter 18. There is also more practice with linear programming.
  - The linear programming chapter is expanded with an additional real application.
- Additional examples and exercises are included for the chapters on mathematical concepts.
  - Additional explanation of functional notation and labeling of coordinate systems is provided.
  - The mathematical concept chapters and the modeling chapters in the last half of the book include more graphs.
  - Several "Earth Notes" about environmental issues are included in the modeling chapters.
  - Things to Do exercises appear at the end of each appropriate section of the mathematical concept chapters.
- Round-off procedures are specified throughout the text.

- Data is updated in tables and figures in the text and in the bibliography of reference sources.
  - An appendix on geometric series is added.
- The Graphing Calculator Manual is omitted.
- Appendix E, Graphing Calculator Programs for *Earth Algebra* is omitted.

DISTINGUISHING FEATURES OF THE TEXT

The text still incorporates the *NCTM Evaluation Standards for School Mathematics* in its use of mathematics to study real-world problems—in particular, global warming and the greenhouse effect. Group work, written and oral reports, modeling using mathematics, and the use of graphing calculators make mathematics a hands-on subject.

Organization of the Text

The first edition is again flexibly organized into five parts. Each part opens with a photo and an overview of the environmental issues to be examined and the mathematical concepts to be used. Concepts traditionally studied in college algebra, such as functions, are covered and then used to create models. Students make predictions, decisions, and recommendations about environmental issues based on these models.

Each chapter that studies environmental issues is preceded by chapters that present the mathematical concepts necessary to complete the study. These "concept" chapters now include exercises at the end of most sections. Each mathematics topic presented is used in an application; there are no extraneous topics.

To complete the environmental applications, the class is divided into groups. Projects for the groups are indicated in the text by the Group Work symbol. Most of these exercises are open ended. The students write papers or present oral reports that summarize the work done for each portion of the application.

Part I of *Earth Algebra* discusses the issue of global warming and how it is affected by carbon dioxide build-up. Students are introduced to the ideas of curve fitting and mathematical models. Part II examines three sources of carbon dioxide emissions: automobiles in the United States,

U.S. energy consumption, and the destruction of tropical rain forests around the world. For each source, three factors are analyzed and then combined to produce functions describing total carbon emissions. In Part III geometric series are discussed and then used to determine total atmospheric accumulation from each emissions source. In the new edition there is a smoother flow among the chapters in Part III plus more figures and text discussion. Part IV shows how two variables—people and money—affect sources of CO_2 emissions. In Part V alternative energy sources are analyzed through linear programming, then student groups devise plans for decreasing future CO_2 emissions. Each group presents its plan to the class.

Instructors can choose which parts are to be covered. The Instructor's Guide contains more information about connections between chapters in *Earth Algebra*.

Appendices at the end of the book cover the variables used in *Earth Algebra*, a table of conversions for units of measurement, and the derivation of the quadratic formula and a brief discussion of complex numbers. Appendix E allows students to check their answers for the odd-numbered Things to Do.

SUPPLEMENT PACKAGE TO ACCOMPANY *EARTH ALGEBRA*

For the Instructor

The **Instructor's Guide** provides background and explanation for the environmental models. It also lists the prerequisite chapters necessary for students to build mathematical models and gives alternative sequences of topics if an instructor wants a more flexible course outline. Alternate syllabi show how to choose topics to change the course focus or for terms of different length. Tips for testing, evaluating Group Work, and classroom management are included. Complete worked-out solutions appear for all the models in the text, and a sample final project is included.

The "Additional Topics to Accompany *Earth Algebra*" section in the Instructor's Guide offers traditional coverage of elementary algebraic functions and conic sections should you need these for your curriculum.

We believe that reformed curriculum is as much about how we teach as what we teach and how we assess it. *Earth Algebra* is not only new in content but an innovative approach to how mathematics is taught. To have the greatest chance at success and student excitement in this course, please see the Instructor's Guide.

The *HarperCollins Test Generator for Mathematics*, available in IBM and Macintosh formats, enables instructors to prepare tests for each of the chapters which present mathematical concepts used in *Earth Algebra*. Instructors may generate tests in multiple-choice or open-response formats, scramble the order of questions while printing, and produce 25 versions of each test. The system features printed graphics and accurate mathematical symbols. The program also allows instructors to choose problems randomly from a section or problem type or to choose questions manually while viewing them on the screen, with the option to regenerate variables. The editing feature allows instructors to customize the chapter data disks by adding their own problems.

GraphExplorer software, available for IBM and Macintosh hardware, allows students to learn through exploration. With this tool-oriented approach to algebra, students can graph rectangular, conic, polar, and parametric equations, zoom, transform functions, and experiment with families of equations. Students have the option to experiment with different solutions, display multiple representations, and print all work.

A set of *computer-assisted tutorials* offers a self-paced, interactive review of concepts in IBM and Macintosh formats. Solutions are given for all examples and exercises as needed.

Transparencies of selected figures and data from *Earth Algebra* are available for instructors. Transparencies of the keyboards for various graphics calculators (Casio, Sharp, and Texas Instruments) are included for classroom use.

For the Student

A **Student's Solution Manual** has worked-out solutions to selected Things to Do.

ACKNOWLEDGMENTS

Key to the development of this text were the users who took the time and effort to give us suggestions and comments. We especially thank the following instructors, as well as the Kennesaw mathematics department, for taking on the effort and the adventure of using early versions in their classrooms:

Tom Adamson, *Phoenix College*
Richard B. Basich, *Lakeland Community College*
Charles Hadlock, *Bentley College*
Lea Campbell, *Lamar University-Port Arthur*
Jay Graening, *University of Arkansas*
Julia Hassett, *DeVry Institute of Technology*
Martha Ann Larkin, *Southern Utah University*
David M. Mathews, *Longwood College*
Don Shriner, *Frostburg State University*
Arthur Sparks, *Georgia Southern University*
Mary Jane Wolfe, *University of Rio Grande*
Ed Zeidman, *Essex Community College*

We also thank these instructors who reviewed the revision of our book:

Laura Corrigan, *Central Florida Community College*
Iris B. Fetta, *Clemson University*
Richard Hickman, *Modesto Junior College*
Linda Horner, *Broward Community College*
Charles Jones, *Ball State University*
Judith M. Jones, *Valencia Community College, East Campus*
Bill Keigher, *Rutgers University*
Donna LaLonde, *Washburn University*
Frances Leach, *Delaware Technical and Community College*
Marveen McCready, *Chemeketa Community College*
Iris McMurtry, *Motlow State Community College*
Charles Wall, *Trident Technical College*

We thank all of our colleagues and staff at Kennesaw State College. In particular we acknowledge and are deeply appreciative to: Pamela Drummond for the project evaluation; Marlene Sims and Stanley Sims for advice in the early development states; Robert Paul for consultation on environmental issues; Marian Fox for suggestions on the manuscript;

student assistants Jennifer Copeland, Pemle Ennis, Debbie Fung-A-Wing, Sherry Hix, Jill Roberts, and Debbie Beck for manuscript preparation; and Dean Herbert L. Davis for his early and continued support. We are especially grateful to our department chair, Tina H. Straley, without whose support this entire project would have been impossible.

We also wish to express thanks to: Jack Pritchard and Anne Kelly for being so venturesome as to publish the preliminary version of this book, and Ed Moura for publishing this edition; Kathy Richmond for her patience and editorial assistance; John Kenelly for his curriculum guidance and calculator expertise; Tom Adamson of Phoenix College for his encouragement and many helpful suggestions, and Ben Fusaro, Chair of the MAA Committee on Mathematics and the Environment, for being a true friend to the earth.

Finally, a special thanks to Pauline, to Lanier, and to Nancy Leigh for their understanding, tolerance, and encouragement.

This project was supported by generous grants from the National Science Foundation and the U.S. Department of Education: Fund for the Improvement of Post-Secondary Education. Initial development was supported by a grant from the Georgia Power Company. For these we are proud and grateful.

Christopher Schaufele
Nancy Zumoff

This work was supported by NSF Grant Number USE–9150624 and U.S. Department of Education: FIPSE Grant Number P116B10601.

To the Student

"If you want to make a course interesting, then you should study something of interest."

M. Sims, March 1990

These simple but deceptively wise words were delivered at the first meeting of an ad hoc committee at Kennesaw State College. The charge to this committee was the following: "Do something about college algebra." Although our Department Chair claims that we volunteered, we, in actuality, were appointed co-chairs of this committee. This appointment was an immediate and direct response to this public statement: "We are wasting our time and the students' time teaching our existing college algebra course."

The opening profundity of Ms. Sims could very well have been the impetus for this book. What, indeed, could be done to make college algebra interesting? And, what makes college algebra boring to practically every student on this planet? The answers to the latter question were easily determined by a review of questions perennially posed by its students. Here are a few familiar ones:

1. "What's this stuff good for?" (in response to most anything);
2. "Who cares?" (in response to thought provoking word problems, such as "Train A leaves New York . . .," or "Sally is twice as old as John . . .");
3. "When will I ever have to do this again in my life?" (in response to simplification of a complex fraction that only Rube Goldberg could have designed);

and lastly, our favorite:

4. "Is x always equal to 2?" (in response to having solved a hideous equation involving roots of rational expressions).

There are answers, of course, to all these. In reverse order:

4. "Yes."
3. "I'm not sure."
2. "*You* should, if you want to pass this course."
1. "Designing electrical circuits," "constructing bridges," "putting a woman on the moon," etc., etc., etc.

The first question probably encompasses all the others (except possibly number 4), so we briefly observe that all the answers provided to this query are true but rather inadequate in yielding a meaningful link between factoring and space walking to any student at the beginning college level.

One of our principal goals, perhaps we should say dreams, is to forever lay question 1 to rest, at least among students of *Earth Algebra*. In an attempt to provide the interest, we have couched college algebra in the vitally important issues of the environment. Neither train A, nor Sally's age, will be of concern herein. Real data about real things are provided in this text, and models are derived to fit. Models use relatively simple algebraic equations, such as linear, quadratic, and rational; also exponential and logarithmic functions are summoned when appropriate. We define a "best" model for a given set of data, and after its derivation, students use it to predict relevant events, then to impose reasonable societal restrictions for improvement of future environmental conditions. The equations are actually used to make decisions about real-world situations. After completion of the first pilot segment of this course, both of us commented that we have been teaching freshman mathematics courses entitled "Decision Math" for some fifteen years now, and this was the first time our students had ever used mathematics to make a decision.

It is our intention that the majority of the material in this text be studied in small groups, ranging from three or four students each. We have found that this stimulates responsibility and confidence. Students' written reports and oral presentations incorporated into the studies place real meaning to "$x = 2$." Two what? Two decades? Two tons of carbon dioxide? And whatever "two" is, how is it relevant?

The text is designed for study with the aid of graphing calculators or appropriate software. These mercifully remove the deep, dark tedium and drudgery involved in manipulation of demonic algebraic expressions.

Earth Algebra is intended for study in a beginning level college algebra course. Most of the standard topics usually covered in a traditional college algebra course are included, although most manipulation of algebraic expressions and graphing of more sophisticated equations are handled with the calculator. However, with the aid of the graphing calculator, we are able to include other topics which may not normally have been covered in the traditional course. Students wishing to enter a standard calculus course may need additional preparation; a supplement covering such topics is included in the Instructor's Guide.

Overall, it is our sincere wish that both the student and the instructor find the reading of this book enjoyable and educational. And to the student: may you learn a little mathematics—and what it's good for—along the way.

Christopher Schaufele
Professor of Mathematics
Kennesaw State Collge

Nancy Zumoff
Professor of Mathematics
Kennesaw State College

PART I

Carbon Dioxide Concentration and Global Warming

Introduction

It was a wintry Thursday afternoon, sunny, but 15° with 30 mph winds, and the furnace was not working in the mathematics building. The professor had just finished teaching her morning calculus class. She gathered up her notes, and instead of returning to her cold office, walked straight through the front doors, and across the campus quadrangle to the parking lot where her car was waiting in the sunlight. She opened the door on the passenger side, slowly sat down in the seat, thumbed through her notes a minute, then comfortably remained there for the next hour preparing for her upcoming analysis class.

What the professor was doing was enjoying the greenhouse effect. Even though it was below freezing, the sun had warmed her closed automobile to a comfortable temperature. And that's what the green-

house effect is: light passes through some enclosing barrier, heat gets trapped within, and it warms the enclosure.

The earth itself is surrounded by a sort of blanket. This blanket is composed of natural gases and traps the heat sent down from the sun, holds it around the surface of our planet, and keeps the air warm. These gases are known as "the greenhouse gases" because the blanket they form works like a greenhouse: it allows sunlight to pass through, but holds in a certain amount of heat.

The proportion of these gases has remained about the same from the end of the last ice age until the beginning of industrial times. As a result, our climate has remained relatively stable. The gases which are the major contributors to the greenhouse effect are, in order of quantity: carbon dioxide (50%), chlorofluorocarbon and other halocarbons (20%), methane (16%), ozone (8%), and nitrous oxide (6%).

Carbon dioxide, CO_2, results from the burning of fossil fuels and deforestation; chlorofluorocarbon, or CFCs, come from air conditioners, refrigerants, foams, and aerosol propellants; methane is produced by swamps, landfills, cattle and other livestock; ozone is a result of fossil fuel burning; and finally, nitrous oxide comes from agricultural chemicals, fertilizer, and also from the burning of fossil fuels.

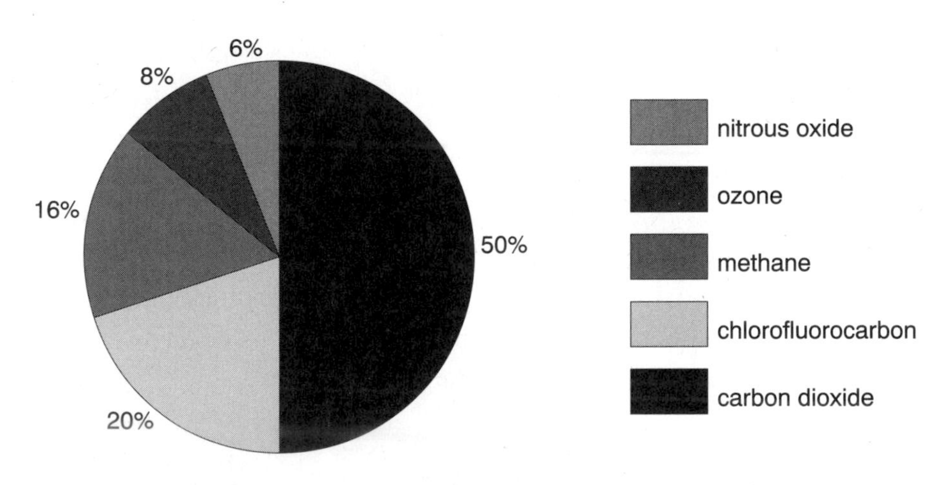

Figure I.1

The climate on our planet is OK—as long as these gases maintain their present proportions. But the world population is increasing rapidly. And all these new people need shelter, need to keep warm or cool, cook food and then eat it, and travel from here to there. Many of these new people want Styrofoam coolers, aerosol spray cans of various products, and the convenience of the myriad of disposable items available at the stores. Then, of course, there need to be lots of new stores, or malls, to sell all these things to all these people. And many of these people use plastic or paper bags to carry their packaged purchases out to their cars. Then there's all the left-over materials, and the no-longer-of-use, or -interest, products which wind up in a landfill.

All these things produce greenhouse gases. So with this increase in production of these gases, it might be suspected that the "normal" proportions present in the atmosphere will be affected. More and more greenhouse gases could make the blanket thicker and thicker, thereby trapping more and more heat at the surface of the earth. And what we get is an increase in average world temperature, which is known as global warming. It is, however, a controversial issue among some scientists as to whether carbon dioxide emission from human activity actually causes global warming.

In the past one hundred years, the average global temperature has increased 1° Fahrenheit. One degree increase over a century may, at first glance, seem insignificant, but note this fact: from the last ice age 18,000 years ago, the earth has warmed by 9° F, or .05° F per century. The present rate of increase is twenty times that of the average rate over the past 180 centuries. That's significant! And, even scarier, as of 1991, the six warmest years in the history of recorded temperature are, in decreasing order of degrees, 1990, 1988, 1983, 1987, 1984, and 1989.

If this trend continues, the eventual effect would be the end of life as we know it on planet earth. One obvious consequence is a change in climates, which would seriously affect farming lands and water supplies. Also, water would expand from the heat and polar ice caps would start melting, causing oceans to rise and cover beaches and other low country. If the global temperature continues to rise at its present rate there will not be time for evolutionary adaptation, thus causing mass extinction of plant and animal species. All these things pose a serious threat to the existence of us humans as well.

Hopefully, most of you are already well aware of the information you have just read, and are doing your environmental duty to save the planet. This book is not intended to provide you with much more information about the causes and effects of global warming. What this book is intended to do is to use concepts from college algebra to study greenhouse gas emission and use these concepts to predict the future. Now this may seem weird to you at this point, that is, how to use mathematics to study about environmental problems. Well, we're going to tell you briefly how it's done—but only briefly—and then we're going to show you. Everybody knows that the only way to learn something is to do it, and that brings to mind the old adage: "Math is not a spectator sport."

In order to study physical or social situations, mathematicians and statisticians build a "model" which looks as much like the situation as possible. The kind of model we use in this book consists of equations and formulae and things of this nature. Here's how we will get these equations and formulae: we have access to historical data which can be plotted on a coordinate system; we find an equation whose graph comes close to all the points determined by the given information. This equation should provide a fairly good approximation to what actually occurred in the past, and can then be used to predict what could be expected to happen in the future, should things continue as they have. Of course, some curves will come closer to the plotted points than others, and we will need to decide which curve is best to use, that is, which equation best approximates the real data. The types of equations we have at our beck and call are linear, quadratic, logarithmic, and exponential; so, given our real data, we first need to decide which type of equation has its graph most shaped like the plotted points. Then we need to determine how to find the best one of these to use, that is, the one which approximates the actual figures the closest.

The procedure described above is called *curve fitting,* and the collection of all of the equations needed to describe a situation is called the *model.*

To the student: the models you will use and build in this course are quite simplistic. This is necessary because of the scope of the course. You should be aware that in reality, many, many other variables, social and political as well as physical, become involved. We have, however, based all models on real data, and most of them prove to be pretty consistent with published predictions.

The mathematical models in *Earth Algebra* are based on current and past trends. Prediction of future events using these models is based on the assumption that these trends will continue. It may not be the case that these trends actually continue; therefore you must understand that your predictions may not come true. Many things can happen in the future which affect these trends, thus changing the outcome.

Also, a model may be accurate for a short period past the present date, but certainly over a long period of time, many other unforeseen events will affect the accuracy of the predictions.

In Part I of this book, you will use the mathematical concepts of functions, linear functions, and composite functions to study atmospheric carbon dioxide concentration and its effect on average global temperature and ocean level.

Coefficients of functions and equations in all environmental chapters should be rounded as follows: linear, two places; quadratic, exponential, and logarithmic, four places. Sometimes we deviate from this practice but not without warning.

Some discrepancy may occur in answers if round-off is done during calculation as opposed to only rounding the final result. In all situations the final answer should be rounded to the same number of places as the original data.

No more discussion—on to the real thing!

CHAPTER 1

Functions

1.1 NOTATION AND DEFINITIONS

This chapter will be a quick and easy look at the mathematical concept of function. For the most part, you'll need to deal with functional notation when you study *Earth Algebra,* so we'll start with that. Almost all of the *Earth Algebra* functions are defined with algebraic expressions, or logarithmic or exponential expressions, or combinations of these. For example,

$$2x + 7$$

defines a function of the *variable* x. If you substitute your favorite number for x and do the arithmetic, you'll get a number for an answer. If your favorite number is $x = 5$, then substitute 5 for x:

$$2(5) + 7 = 17$$

is the answer.

A really "loose" way to describe a function is this: a function assigns to each number x a unique number y.

In the above example, the function defined by $2x + 7$ assigns to the number $x = 5$ the number $y = 17$. It assigns to the number $x = 0$ the number $y = 7$, and so on.

Note: Chapter 1 is a prerequisite for Chapters 2 and 3.

Just like people, functions need names. Name the one defined by $2x + 7$ by the letter f. Now f is a function of the variable x, so functional notation is

$$f(x) = 2x + 7.$$

Read the notation on the left of the equal sign: "f of x." It works like this: if you substitute 5 for x, then you write

$$f(5) = 2(5) + 7 = 17,$$

or if $x = 0$,

$$f(0) = 2(0) + 7 = 7.$$

You say for these operations: "$f(x)$ is evaluated at 5," or "at 0," or, abbreviated, "f of 5" or "f of 0."

Whatever you substitute for x goes in the parentheses in the functional notation: $f(-1) = 2(-1) + 7 = 5$. If you substitute a cat for x, then

$$f(\quad) = 2(\quad) + 7.$$

Got it? Some people think that $f(x)$ means multiply f times x, but it doesn't. "f" is the name of the function and x is its variable.

A couple more examples of functions are:

1. $f(x) = x^2 - 3x + 1.$
2. $g(x) = \dfrac{1}{x - 3}.$

Evaluate $f(x)$ at 2: $f(2) = (2)^2 - 3(2) + 1 = -1$; evaluate $g(x)$ at 7: $g(7) = \dfrac{1}{7 - 3} = \dfrac{1}{4}.$

It isn't hard to evaluate functions on your calculator, but each kind does it a little differently. First you have to tell the calculator what your function is, that is, you must enter the expression for your function (which might be called $y_1, y_2, \ldots,$ or $f_1, f_2, \ldots$). Then assign the value to x, press the key for your function and enter. For example, to evaluate $2x + 3$ at -5, you would enter the expression

$$2x + 3$$

for the function y_1 (or f_1), assign -5 to x, press y_1 (or f_1), and enter. Your screen will show -7. To find out more, read your manual, or ask your instructor or a knowledgeable friend.

1.2 DOMAINS AND RANGES

Try this: evaluate $g(x) = \dfrac{1}{x - 3}$ at 3. Big problem. Division by zero is illegal. This means that you can't just substitute any number into a function. The function $g(x)$ is not defined when $x = 3$, so you may have to restrict the numbers which can be used to evaluate a particular function. The only real problem with evaluating $g(x)$ occurs when the denominator equals 0, that is, when $x = 3$, so $g(x)$ is defined for all numbers except 3. Usually, the domain of a function consists of all numbers at which the function is defined. Thus the domain of $g(x)$ consists of all numbers except $x = 3$.

What about the domain of $f(x) = x^2 - 3x + 1$? No problem here with substituting any x, so its domain consists of all numbers.

A kind of different situation is presented by the function

$$h(x) = \sqrt{x}.$$

No negative numbers can be substituted for x because you can't take the square root of a negative number and get a real number. (In *Earth Algebra,* imaginary numbers will not be considered. See Appendix C for imaginary numbers, or complex numbers.) But, anything else is OK to substitute in for x, so the domain of $h(x)$ is the set of all real numbers $x \geq 0$. See what happens if you try to evalute this function at $x = -3$ using your calculator.

There are also practical considerations to take into account when worrying about the domain of a function. All of the functions of *Earth Algebra* have practical purpose in their lives. For example, suppose that $F(w)$ defines the number F of catfish which live in a certain north Georgia lake as a function of the number w of gallons of toxic waste dumped into a creek upstream of the lake by a certain chemical factory. ($F(w)$ could be defined by some sort of mathematical expression, but that doesn't matter now. But whatever defines it, you can bet that it is decreasing; that is, the larger w is, the smaller F gets!) The variable w counts gallons of toxic waste so it would make no sense to say $w = -10$, right? "Negative ten gallons of waste were dumped today." Your friends would think you are crazy. So the domain of $F(w)$ would be all: $w \geq 0$.

There are two things to consider when determining the domain of a function:

1. At which numbers is the mathematical expression defined?
2. What are the practical limitations?

EXAMPLE 1.1

Determine the domain of $f(x) = 2x + 1$.

Anything can be substituted for x. Domain = all real numbers. ▲

EXAMPLE 1.2

Determine the domain of $f(x) = \sqrt{x - 1}$.

Whatever's under the radical must be nonnegative; that is, $x - 1 \geq 0$; thus $x \geq 1$. Domain = all real numbers $x \geq 1$. ▲

EXAMPLE 1.3

Suppose $J(p)$ defines the unemployment rate J (for jobs) as a function of gross national product p. The gross national product is another counting variable, it is the monetary value of all goods produced in the country. Hence $p \geq 0$, and the domain of $J(p)$ would be all real numbers $p \geq 0$. ▲

Up until this point, we've only been talking about the numbers which can be substituted into the function (domain), but what about the answers you get after you substitute? All the answers that you get when you substitute all the domain numbers into the function comprise what is known as the range of the function.

If $f(x) = x^2$, then $f(2) = 4, f(-1) = 1, f(7) = 49$, and so on. So the numbers 4, 1, and 49 are all in the range of $f(x)$; of course these are not all the range numbers; you could plug in numbers all day and never get all the range of $f(x)$. There's a nicer way to see what the range is graphically; we'll talk about that later. But do note this now: -4 is not in the range because $x^2 \geq 0$ always. As a matter of fact, no negative number is in the range.

This is enough about range for now—more later.

1.3 GRAPHS OF FUNCTIONS

We've heard lots of students say that they hate graphs, but they shouldn't. Graphs are useful; they are pictures of functions. Sometimes by graphing a function it is easier to see certain information that a function provides.

If $f(x)$ is your function, then first set $y = f(x)$. If $f(x)$ is defined by some expression, then what you now have is an equation in two variables x and y. x is called the *independent variable* (because that's the one you substitute the numbers for), and y is called the *dependent variable* (because y is the answer and it depends on what x is). The graph of the function $f(x)$ is sketched in the plane formed by an (x, y) coordinate system. This consists of a horizontal axis for x, or whatever the independent variable is, and a vertical axis for y, or whatever the dependent variable is. Each of these axes is a copy of the real line; they intersect at zero on each, positive numbers are to the right and up; negative numbers are to the left and down. Points in this plane are located by a pair of numbers (a, b); the first one, a, comes from the horizontal axis, and the second, b, comes from the vertical axis. See Figure 1.1. If $x = a$, and $b = f(a)$, then (a, b) is a point on the graph of $f(x)$. The *graph* of $f(x)$ consists of all points (a, b) such that $b = f(a)$, and a varies over the entire domain.

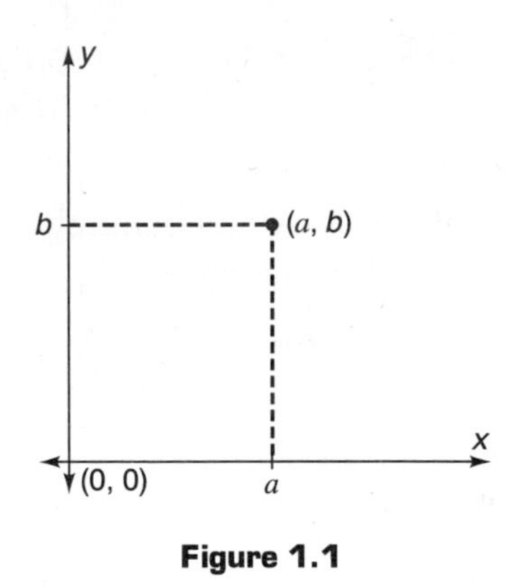

Figure 1.1

Remember when you first learned how to graph a function, or equation, you'd substitute three or four numbers in for x, find the corresponding y value, plot your points, and connect the dots? This may not give a very good picture of your function because you left out so many domain numbers. You ask how could anyone possibly substitute every one of that infinity of numbers in the domain? Of course no one can. But there is a way to see the graph by using your calculator. As you study each type of function in *Earth Algebra,* we'll tell you about its particular intricacies—what's important about each—then discuss how to see these important things on your calculator screen. Graphing with the calculator can be fun. And there'll be plenty for you to practice with when the time comes.

One final note on functional notation: none of the functions of *Earth Algebra* use the variables x and y. They are named according to the purpose they serve. For example, suppose you wanted to write a function which could predict the number of acres of rain forest which would be remaining on this planet in any given year. What would you name your function? How about RF? And what would RF be a function of? How about t (for time)? So your function would be $RF(t)$ where RF = number of acres of rain forest in year t. And then your graph would be on a (t, RF) coordinate system. (We adopt a fairly common practice in mathematics of denoting a function and its functional value by the same letter.) See how that works?

Finally, here is some practice for you.

THINGS TO DO (ALIAS EXERCISES)

For Exercises 1–10, evaluate the given function at the indicated values, and find the largest possible domain for each. Try evaluating the first three functions "by hand," and for the others, let your calculator do the work.

1. $f(x) = 5 - 2x$; evaluate $f(x)$ at 2, -3, 1.01, and $-\frac{1}{2}$.

2. $g(x) = 2x^2 - 5x + 17$; evaluate $g(x)$ at 2, -3, 1.01, and $-\frac{1}{2}$.

3. $h(x) = \dfrac{3x^2 - 5}{4x - 7}$; evaluate $h(x)$ at 2, -3, 1.01, and $-\frac{1}{2}$.

4. $F(t) = 517 - 2.14t$; evaluate $F(t)$ at 2, -3.15, 0.000051, $\frac{7}{5}$, and $-\frac{1}{2}$.

5. $T(x) = .005x - 120$; evaluate at $x = 2.4$, 24, 240, 2400, and 24,000.

6. $G(x) = .002x^2 - 5.13x + 729$; evaluate $G(x)$ at 2, -31, 121.201, $\frac{2}{3}$, and $-\frac{1}{2}$.

7. $H(x) = \dfrac{3x^2 - 5}{2x - 11}$; evaluate $H(x)$ at 2, -3.3, 11.001, and 5.5.

8. $S(u) = \dfrac{(.005u - 20.2u^2 + 9.1)(10.9 + 2.3u)}{12u - 2.1}$; evaluate $S(u)$ at 0, 50, 100, and 5000. Try this on your calculator: enter each of the parenthetical expressions in Exercise 8 as a y_1, y_2, and y_3, then let $y_4 = S(u)$, the appropriate combination of the three pieces.

9. $R(s) = \sqrt{63 - 2.1s}$; evaluate at $s = 0, -2, 30, -1.52$, and 50.

10. $P(x) = 150{,}000x(12 - .0005x)$; evaluate at $x = 10, 100, .001$, and 108.

Linear Functions

2.1 SLOPE-INTERCEPT EQUATION

The graph of a linear function is a straight line. A linear function looks like $f(x) = mx + b$, where m and b are constants. These constants are very significant: the number m is the *slope* of the line, and the point $(0, b)$ is its *y-intercept*. (The y-intercept of any graph is the place where it crosses the y-axis, and the *x-intercept* is the place where it crosses the x-axis.)

The y-intercept is found by setting $x = 0$ (any point on the y-axis has $x = 0$) and then determining what y is.

If $x = 0$, then $f(0) = m \cdot 0 + b = b$, so $y = b$ and $(0, b)$ is the y-intercept, just as we said!

Now, to see the significance of the slope, set $x = 1$ and substitute:

$$f(1) = m \cdot 1 + b = m + b.$$

Note: Chapter 2 is a prerequisite for Chapter 3.

You have this information:

$$\text{when } x = 0, \quad y = b;$$

$$\text{when } x = 1, \quad y = m + b.$$

See Figure 2.1.

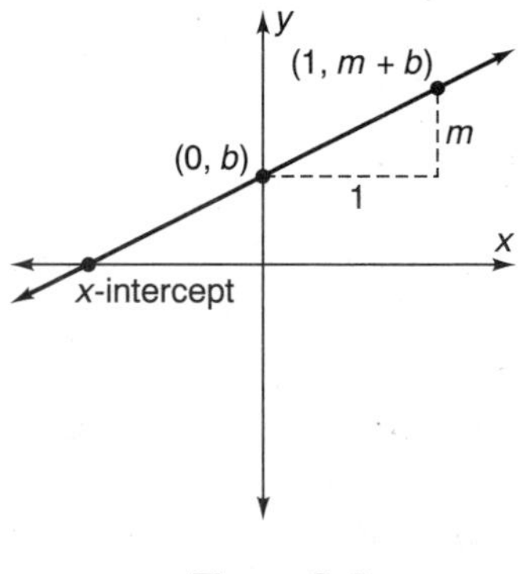

Figure 2.1

This says that a one-unit change in x (from 0 to 1) results in an m-unit change in y (from b to $m + b$). The horizontal axis is the x-axis, so any change in x represents a horizontal change; the vertical axis is for y, so changes in y are vertical changes. A one-unit horizontal change in the graph of $y = mx + b$ results in an m-unit vertical change. This is why m is called the *slope* of the line; it tells how steep it is. If you look at any two points on your line, and measure the horizontal change and the vertical change from one point to the other, then this slope is

$$m = (\text{vertical change}) \div (\text{horizontal change}).$$

Look back at the original function

$$f(x) = mx + b,$$

or the equation

$$y = mx + b.$$

This is called the *slope-intercept equation* for a line (here's a question: why?). This is a nice equation because it's simple and gives lots of information about the line.

Some particular examples are below.

EXAMPLE 2.1

Determine the slope and y-intercept of $y = 2x + 1$.

$$\text{Slope} = 2, \quad y\text{-intercept} = (0, 1).$$

See Figure 2.2. ▲

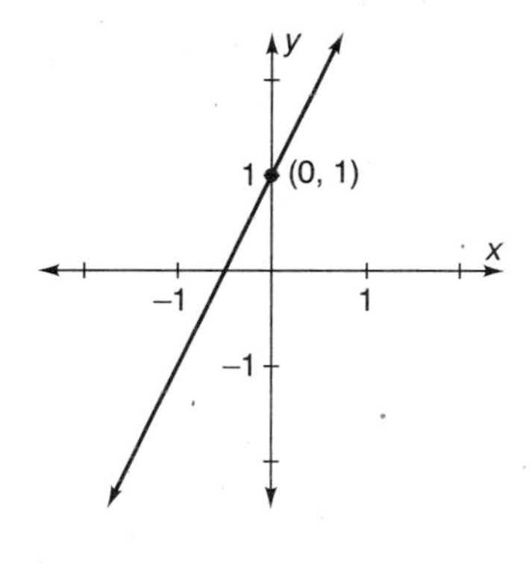

Figure 2.2

EXAMPLE 2.2

Determine the slope and y-intercept of $y - 1 = -2(x - 3) + 7$.

First write in slope-intercept form.

$$y = -2x + 14.$$

Slope $= -2$, y-intercept $= (0, 14)$. See Figure 2.3. ▲

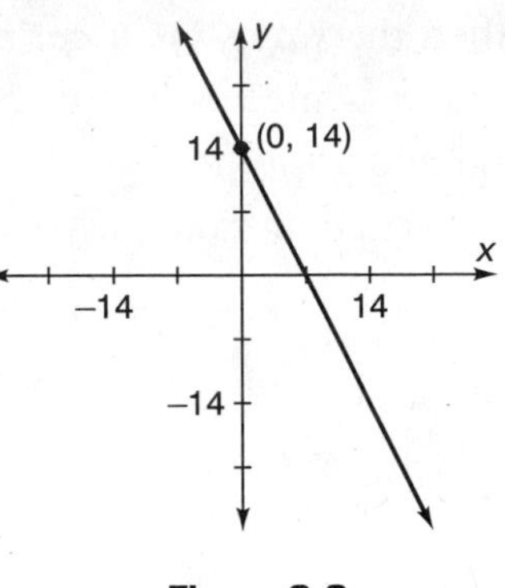

Figure 2.3

EXAMPLE 2.3

Determine the slope and y-intercept of $2y = 3x + 1$.

Careful, the slope is not 3 nor is the y-intercept (0, 1). Write in slope-intercept form; that is, solve for y:

$$y = \frac{3}{2}x + \frac{1}{2}.$$

Now the slope is the coefficient of x, and the y-intercept is the constant at the end.

$$\text{Slope} = \frac{3}{2}, \quad y\text{-intercept} = \left(0, \frac{1}{2}\right).$$ See Figure 2.4. ▲

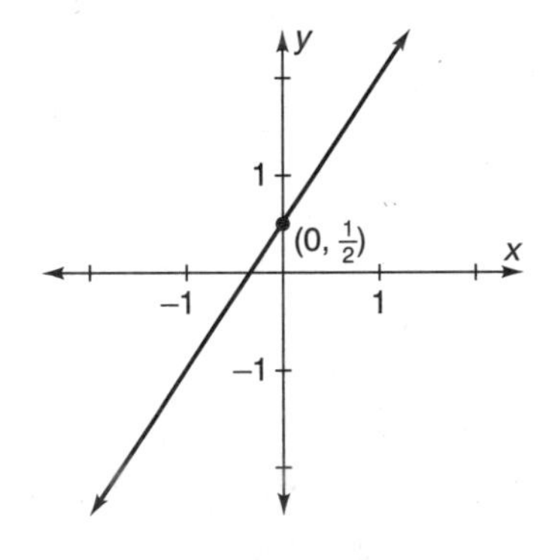

Figure 2.4

2.1 THINGS TO DO

Write the following equations in slope-intercept form, and determine the slope and the y-intercept.

1. $y = -1.4x$
2. $y + 2 = 3x - 5$
3. $y + 1.3 = -2(x - 1.1)$
4. $2(x - 4) = 3(y + 7)$
5. $3x + 5y = 18$
6. $2x = 5 - 3y$
7. $12(2x - 1) - 5(3y + 2) = 8$
8. $1.2(3x + 5y) - 3.2(3x - 5) = 0$
9. $2(2x + y) - 4(x + y) = 6$
10. $5(x + 2) - 3(y + 1) = 5x - 3$

2.2 SLOPES

Let's get numerical! First, remember that the slope is the ratio of the vertical change to the horizontal change between two points on the graph of a line. If the two points are (x_1, y_1) and (x_2, y_2), the horizontal change is $x_2 - x_1$ and the vertical change is $y_2 - y_1$. Hence the formula

$$m = \frac{y_2 - y_1}{x_2 - x_1}$$

gives you a way to determine the slope of a line if you know two points on that line. See Figure 2.5.

What if you needed to know the slope of the line which goes through the points (1, 2) and (3, 4)? Having mastered college algebra, you could very quickly answer, "It's quite simple. The slope is 1." But then suppose

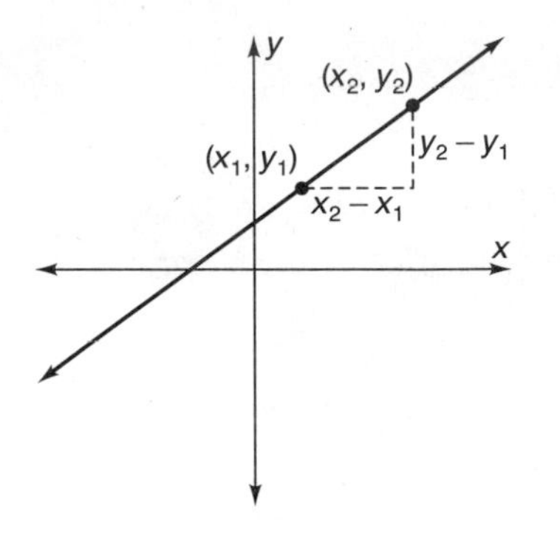

Figure 2.5

"minuses" got involved and the points changed to $(-1, 2)$ and $(3, -4)$? But negatives wouldn't bother you in the least. You reply, "That slope is

$$m = \frac{(-4) - 2}{3 - (-1)},$$

which is $\frac{-6}{4}$, and that's $-\frac{3}{2}$."

There's an easy way to get the graph of a linear function using its slope. It's what we call "Doing the three-second graph." Suppose you know a point on a line and the slope of that line. Then think of slope in terms of ratio of vertical change to horizontal change to draw this graph. The next two examples illustrate this concept best.

EXAMPLE 2.4

Graph the line which passes through (2, 5) and has slope 4.

Write $m = 4$ as a fraction,

$$m = 4 = \frac{4}{1};$$

then the vertical change is 4 and the horizontal change is 1. Your starting point is (2, 5). Move vertically 4 units up, then horizontally 1 unit to the right. This determines another point on this line; that point is (3, 9). You can count points on the coordinate system to get this point or you can add to get $(2 + 1, 5 + 4)$. Now, simply draw the line through these points. See Figure 2.6. ▲

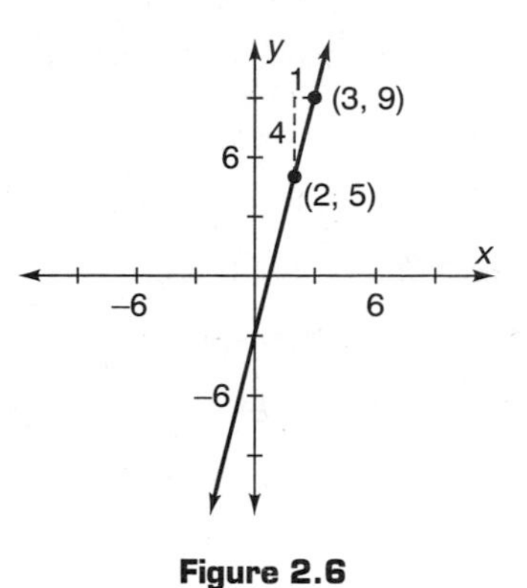

Figure 2.6

EXAMPLE 2.5

Graph the line which passes through (1, 2) and has slope $-\frac{3}{5}$.

Think of the slope as being -3 divided by 5; write the slope as

$$m = \frac{-3}{5},$$

so the horizontal change is 5 and the vertical change is -3. The negative in front of the 3 means that your vertical change is downward. Start at (1, 2), go down 3 units, then right 5 units. The new point is $(1 + 5, 2 - 3) = (6, -1)$. Draw the line through (1, 2) and $(6, -1)$ and you've got it! See Figure 2.7. ▲

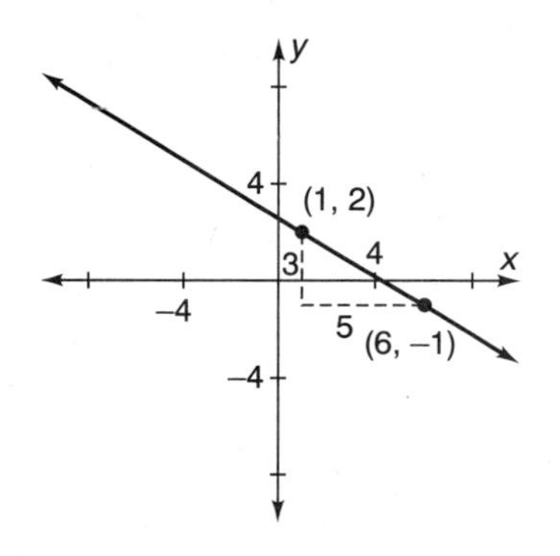

Figure 2.7

Before we move on, here are some useful pieces of information about slopes.

- If $m > 0$ the linear function is increasing, that is, the graph is rising from left to right.
- If $m < 0$, then the linear function is decreasing, or its graph is falling as you go from left to right.
- If $m = 0$, then the line is horizontal; the linear function is constant.

See Figure 2.8.

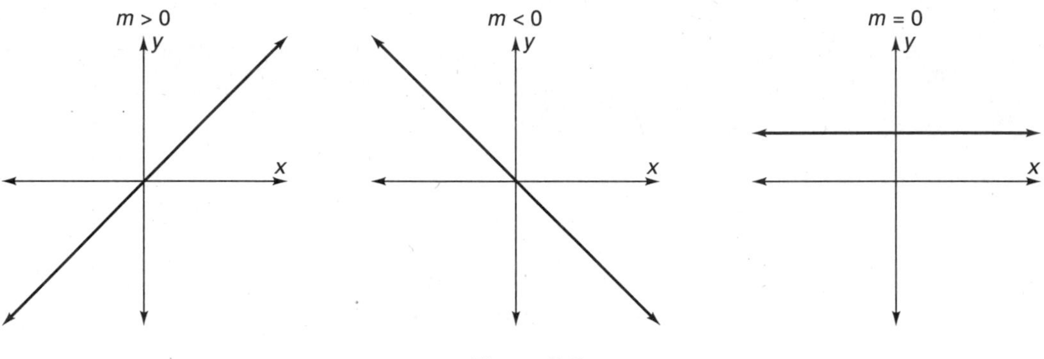

Figure 2.8

There's this one other thing about slopes and lines which you won't be concerned much about in *Earth Algebra,* but we'll tell you anyway—just so your education is complete: vertical lines have no slope—their slopes are undefined. Since slope is

(vertical change) $\div$ (horizontal change),

a vertical line has zero horizontal change; so you'd have to divide by zero, which is undefined. Finally, the equation of a vertical line looks like $x = a$, where a is constant. See Figure 2.9.

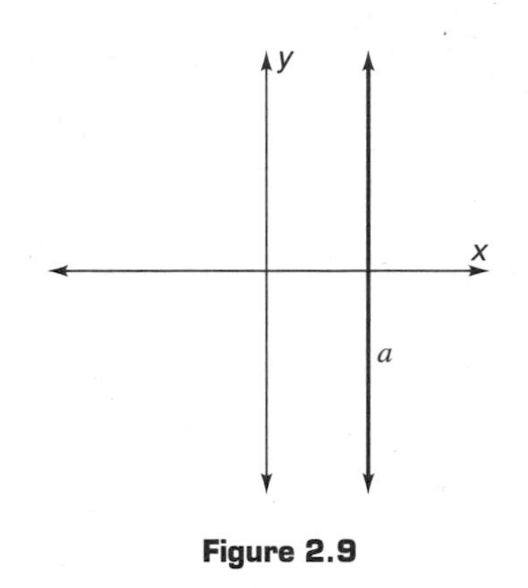

Figure 2.9

2.2 THINGS TO DO

Use the "three-second" technique to graph the line through the given point with the given slope.

1. $(2, -3), \quad m = 4$
2. $(1.7, 2.1), \quad m = -2$
3. $(2, 5), \quad m = \frac{-3}{5}$
4. $(5, 12), \quad m = \frac{8}{3}$
5. $(2.3, 5), \quad m = 0$
6. $(5, 2), \quad m = \frac{2}{5}$
7. $(-3, -1), \quad m = -1$
8. $(3.2, 1.5), \quad m = 2$

2.3 GRAPHS

Linear functions have, of course, the easiest kind of graph. All you need are two points, and any two points uniquely determine a line. If you've carefully read the sections which come before this, you already know one way to graph a linear function, the slope-intercept way: first, write your linear equation in the form $y = mx + b$; then start at the y-intercept and think of the slope as

(vertical change) $\div$ (horizontal change).

An example is:

$$y = -2x + 1. \left(\text{Note: } m = -\frac{2}{1}.\right)$$

Start at the y-intercept $(0, 1)$, move right 1 unit, then move down 2 units to get your second point $(1, -1)$ and draw the line. See Figure 2.10.

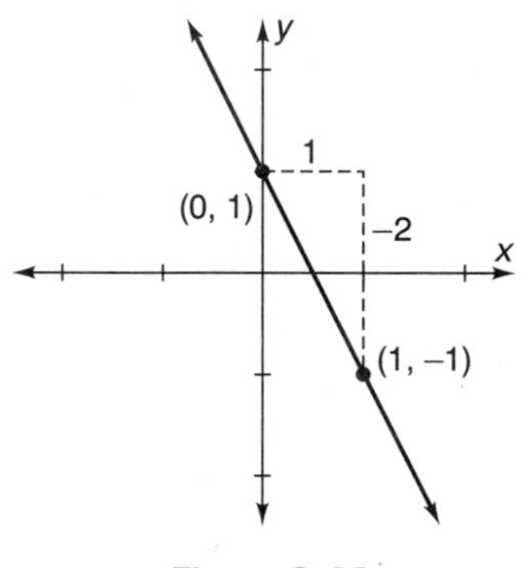

Figure 2.10

Another way to graph a line is to determine both the y-intercept and the x-intercept. The y-intercept is easy to determine when the equation is in slope intercept form $y = mx + b$; it's $(0, b)$. But to get the x-intercept, you have to work a little—not much, though. What you have to do is to solve the linear equation

$$0 = mx + b$$

(this is a result of setting $y = 0$). An example of this follows.

EXAMPLE 2.6

Graph the line $y = 3x + 6$ by finding its x- and y-intercepts.

The y-intercept is (0, 6). For the x-intercept, take $y = 0$ and solve $0 = 3x + 6$

$$3x = -6 \qquad \text{and} \qquad x = -2$$

so the x-intercept is $(-2, 0)$. Plot these points and draw your line. See Figure 2.11. ▲

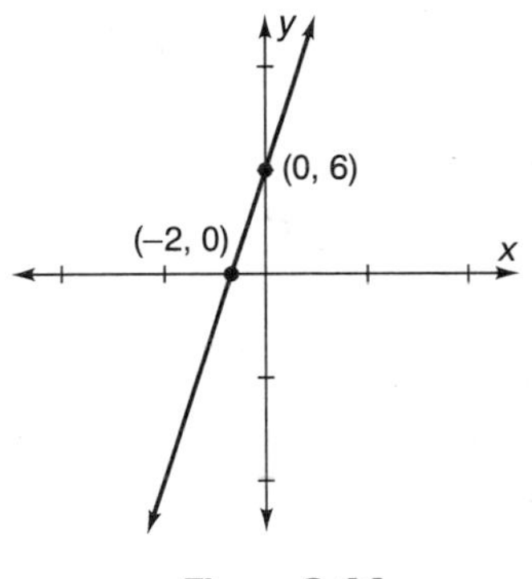

Figure 2.11

There may be a problem with graphing by using the two intercepts; there may not be two. Look at this next example.

EXAMPLE 2.7

Graph the line $y = 2x$.

When $x = 0$, then $y = 0$, and vice versa, so the x-intercept and the y-intercept are both the origin (0, 0). In this case, to get another number you can substitute your favorite number (as long as it's not zero) for x and find y. If -7 happens to be your favorite number, then

$$y = 2(-7) = -14$$

and your second point is $(-7, -14)$. Plot these points and draw your line. See Figure 2.12. ▲

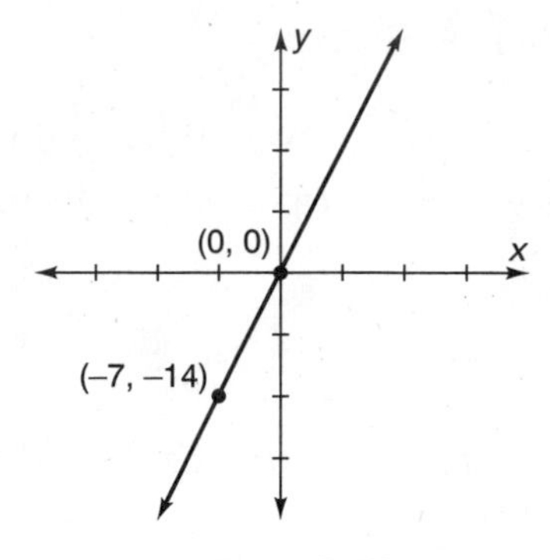

Figure 2.12

2.3 THINGS TO DO

Graph each of the lines below. Label the x-intercept and the y-intercept, and find the slope.

1. $2x + 3y = 12$
2. $y = 1 - 4x$
3. $y = 1.5x$
4. $x = 7$
5. $y + 3 = 1$
6. $y = -3x$
7. $2(x - 1) + 3(1 - y) = 6$
8. $x + 3y = -2x + 3y - 5$

2.4 GRAPHING LINES WITH YOUR CALCULATOR

It's easy to graph linear equations with your calculator, but first, you must write your equation in the slope-intercept form

$$y = mx + b.$$

Then you can graph. We'll start with an easy example.

EXAMPLE 2.8

Graph $2x + 3y = 9$ using your calculator.

In slope-intercept form, it is

$$y = -\frac{2}{3}x + 3.$$

Enter the function in the calculator and press the **graph** key. You probably will see something like Figure 2.13. If your calculator graph doesn't look exactly like this picture, it's probably because your x and y ranges are set differently. In Figure 2.13, both x and y ranges are set from -10 to $+10$.

$$x \text{ min} = -10,$$
$$x \text{ max} = 10,$$
$$y \text{ min} = -10,$$
$$y \text{ max} = 10.$$

If you reset your ranges like this, then you will see exactly the same picture as Figure 2.13. To accomplish this press the **range** key, and enter these numbers. ▲

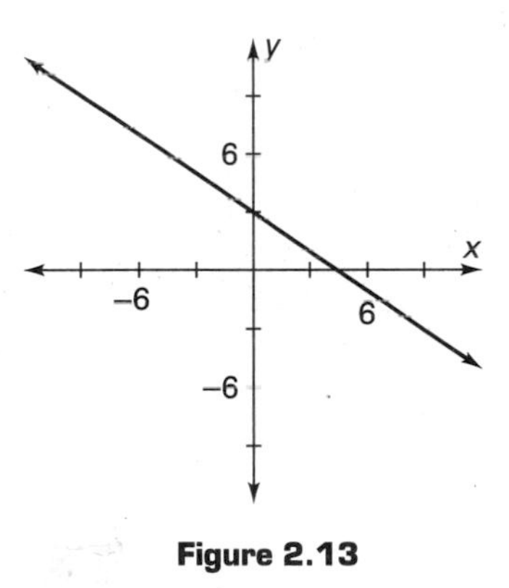

Figure 2.13

EXAMPLE 2.9

Graph $y = 10{,}000 - .02x$ on your calculator and show intercepts.

Enter the function and press **graph.** See anything other than axes? Your answer is "No!" unless you changed your ranges from the previous settings. Think about it: the y-intercept is 10,000, which is considerably larger than 10, and the x-intercept is 500,000, also bigger than 10. So this line cuts across the plane from (0, 10,000) to (500,000, 0), completely missing your screen! In order to see relevant information (intercepts) you should set x min < 0, x max $> 500{,}000$ and y min < 0, y max $> 10{,}000$. We choose

$$\begin{aligned} x \text{ min} &= -100{,}000 \\ x \text{ max} &= 600{,}000, \\ y \text{ min} &= -500, \\ y \text{ max} &= 12{,}000. \end{aligned}$$

Figure 2.14 shows the graph we get on our calculator with these ranges. Note that these range settings are somewhat arbitrary and could be different—as long as relevant information is displayed (in this example, intercepts). ▲

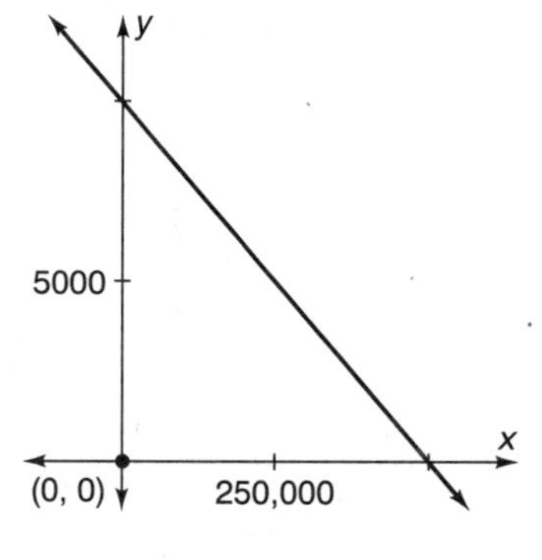

Figure 2.14

EXAMPLE 2.10

Suppose the function $P(t) = 2.374t - 4473.74$ defines the population P (in millions) of the United States in year $t \geq 1940$. For example, in 1950, the population was (substitute $t = 1950$)

$$P(1950) = 155.56 \text{ (in millions)},$$

or 155,560,000 people.

Graph this function on your calculator so that the graph shows the population for years 1940–2000.

Enter this function; if your calculator only graphs using variables x and y, you must use x for t, and y for P. Next, set x- and y-ranges. In this example, relevant information is contained between the years 1940 and 2000. So set the x-range to be

$$x \text{ min} = 1940,$$
$$x \text{ max} = 2000.$$

That was easy, but what about the y-range? First, note that the slope of this line is positive, so the function is increasing. This means the population for this time period was least in 1940, and will be greatest in 2000. In $t = 1940$,

$$P(1940) = 131.82$$

and in $t = 2000$,

$$P(2000) = 274.26.$$

Now set your y-range:

$$y \text{ min} = 130, \quad y \text{ max} = 275,$$

and graph. Figure 2.15 shows this graph. ▲

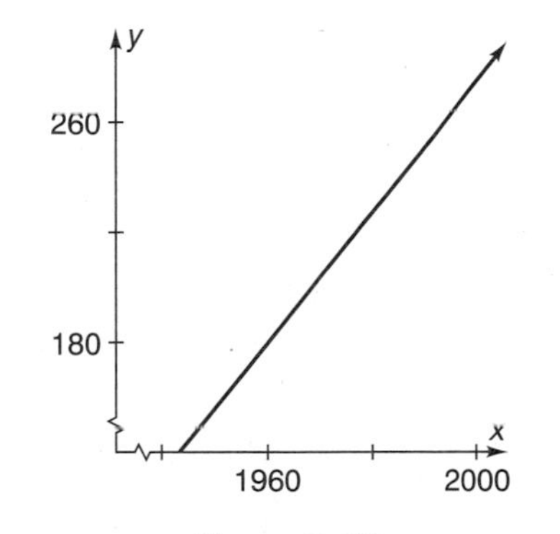

Figure 2.15

These calculator techniques are important, but you'll still need to know the slope and intercept for other reasons, so you'll need to learn them also.

2.4 THINGS TO DO

Graph each of the functions below using your calculator. Set the viewing rectangle to show the x-intercept and the y-intercept; find the slope.

1. $f(x) = 2.4x + 12$

2. $g(x) = 9 - .02x$

3. $CO(t) = .0003t + 12$

4. $r(t) = -102t + 17$

5. $m(t) = 2(3t - 5) + .1(1 - 2t)$

6. $T(x) = 1.8(x - 100) + 18$

7. $E(t) = .01t + 115.2$

8. $A(x) = 10{,}000(2x - 1) + 3000$

2.5 WRITING YOUR OWN EQUATIONS

Until now, linear functions have appeared in front of you for you to graph or analyze. But the main thing you need to do for *Earth Algebra* is to make your own functions and equations. In order to write the equation of a line, you need certain information. In particular, either one of these two pieces of information will suffice:

1. one point on the line, and its slope; or
2. two points on the line.

Once you have either 1 or 2, you need to know what to do with it. This leads us to another type of equation for a line, so we'll discuss that.

Remember the slope formula

$$m = \frac{y_2 - y_1}{x_2 - x_1},$$

where (x_1, y_1) and (x_2, y_2) can be any two points on the line. Modify this slightly—hold (x_1, y_1) fixed (a known point), and let (x_2, y_2) be a variable point, and so as not to confuse it with a fixed point, leave off the subscripts and just call it (x, y). Now, multiply both sides of the equation $m = \frac{y - y_1}{x - x_1}$ by $x - x_1$, to get this nice equation

$$y - y_1 = m(x - x_1).$$

The reason this equation is so nice is that you can use it to write equations of lines that you need to study environmental problems. It is called the *point-slope equation for a line,* because if you look at it carefully, you can see a point on the line and the slope of the line. Those are, respectively, (x_1, y_1) and m.

If the equation,

$$y - 3 = 4(x - 5)$$

should appear before you, you look closely and see that (5, 3) is a point on the line, and that 4 is its slope!

If this equation fades from view and the new one,

$$y + 7 = 3(x - 1),$$

materializes, you know that $(1, -7)$ is on this line, and the slope is 3.

Back to equation writing. Here's how to use the point-slope equation to write the equation of a particular line. Take the two cases, one by one.

EXAMPLE 2.11

Write the equation of the line which passes through $(-1, 3)$ and has slope 5.

Substitute $x_1 = -1$, $y_1 = 3$ and $m = 5$ to get

$$y - 3 = 5(x - (-1)).$$

Simplify and solve for y, and the equation is

$$y = 5x + 8.$$

Easy! ▲

Suppose you know two points. This time, you must do a tiny bit of work before substituting—you must compute the slope using the slope formula, then substitute either of your known points for (x_1, y_1) and your newfound slope m. Here's an example.

EXAMPLE 2.12

Suppose you know that the two points $(2, -1)$ and $(-4, 3)$ are on your unknown line. Write the equation of this line.

First, find m:

$$m = \frac{-1 - 3}{2 - (-4)} = \frac{-4}{6} = -\frac{2}{3}.$$

Then use $x_1 = -4$, $y_1 = 3$ (or use the other point if you'd rather, you get the same answer) to get

$$y - 3 = -\frac{2}{3}(x - (-4)).$$

Simplify to

$$y - 3 = -\frac{2}{3}x - \frac{8}{3},$$

solve for y, and get

$$y = -\frac{2}{3}x + \frac{1}{3}.$$

If you prefer functional notation,

$$f(x) = -\frac{2}{3}x + \frac{1}{3}$$

is your answer. ▲

2.5 THINGS TO DO

Write the equation for the line satisfying the given information, then write it in slope-intercept form. Graph the equation, and show both of the intercepts.

1. Passes through the points (3, 5) and (2, 1).

2. Passes through the points (2, −1) and (−1, 2).

3. Passes through the points (3, 1) and (−2, 1).

4. Passes through the point (2, 4) with slope −1.

5. Passes through the point (−3, 2) with slope $\frac{1}{2}$.

6. Passes through the points (10, 42.3) and (20, 51.5).

7. Passes through the point (0, −12) with slope 16.4.

8. Passes through the points (100, 2.2) and (150, 1.9).

9. Passes through the points (−30, 280) and (51, 350).

10. Has slope −.02, passes through the point (125, 21.3).

11. Passes through the point (12, 2.4) with slope 0.

12. Passes through the points (−2, 5) and (2, −5).

13. Passes through the points (20, 18.5) and (48, 19.4).

14. Passes through the points (.0035, 2) and (.0012, 5).

CHAPTER 3

Atmospheric Carbon Dioxide Concentration

3.1 YOUR FIRST MODEL

There's more carbon dioxide than any other greenhouse gas. It is responsible for approximately fifty percent of the greenhouse effect. There's always been CO_2 in the atmosphere, and that's good—to a certain extent. But the concentration has recently been increasing significantly. Since the industrial revolution human activity which emits carbon dioxide has also significantly increased. Whether or not this is the cause of increased atmospheric concentrations is a controversial issue. Perhaps you can make decisions of your own after studying *Earth Algebra.* Here are just a few things that are most likely to be a part of your life which cause CO_2 to be emitted into the atmosphere: riding in your car to the mall (or anywere); reading under the electric light; sitting in someone's teak chair; eating a hamburger at some fast-food restaurant; cranking up the furnace on a cold winter day, or the air conditioner on a hot summer day. The big sources of carbon dioxide are energy production and deforestation, both of which are for human consumption. We'll take a close look at some of these sources later, but now, for your first adventure in mathematical modeling, we look at atmospheric carbon dioxide concentration. In Table 3.1 the data given is the CO_2 concentration in parts per million (ppm) by year. (Exercise: For the year 1980, put one million sky-blue dots on a piece of paper, then near the top, make three hundred thirty-eight and one half of them dirty.)

These data were compiled at the Mauna Loa Observatory in Hawaii.

TABLE 3.1

Year	CO_2 Concentration (in ppm)
1965	319.9
1970	325.3
1980	338.5
1990	354.0

Source: World Resources Institute, World Resources, 1992–93.

Our goal is to find a function which best approximates these data. By this we mean the following: first the data can be represented by points in the plane, and the graph of the approximating function should lie very close to all the points. Once this approximating function has been determined, we can use it to predict future events. These predictions are based on past trends which are reflected by the data and on the assumption that the trends will continue. However it should be clear to you that there are reasonable limitations on the time over which predictions can be made; too many things change over time. Other limitations will become obvious to you as you work through *Earth Algebra.*

The determination of the approximating function is called *curve fitting,* and the function together with all its restrictions comprises a *mathematical model.* The types of functions we use in *Earth Algebra* are linear, quadratic, exponential, and logarithmic; the choice is dependent upon the pattern of the plotted points. Figure 3.1 shows the general forms for the graphs of these functions.

Figure 3.1

These forms can also be turned upside down when needed. At this point, you have already studied linear functions; in later chapters, you can learn more about the other functions.

Statisticians and mathematicians can use more sophisticated types of functions and methods to approximate data, but these functions and methods are beyond the scope of a college algebra course. Also, our models will be very simple; in real life many, many variables need to be considered and this would involve more advanced techniques. Even with the best, most advanced models, external events may occur which cause the model to become totally inaccurate. What we provide you with in *Earth Algebra* is an idea of the flavor of using mathematics as a tool for prediction and decision making.

Now, here's a step-by-step procedure for modeling the data in Table 3.1.

Step 1. Make points out of the data. The points have as first coordinate "year," and as second coordinate "CO_2 in ppm." In order to make the numbers more manageable, adjust the years in the first coordinate so that the year 1965 corresponds to 0; in other words, if t is the variable representing time, than $t = 0$ means 1965, and in general, $t =$ number of years after 1965. Also, let CO_2C denote the variable for the second coordinate; that is, $CO_2C =$ carbon dioxide concentration in ppm. Functional notation would be this:

$$CO_2C(t) = \mathrm{CO_2 \text{ concentration in year } 1965} + t.$$

For example in 1990, CO_2 concentration was 354.0 ppm. This translates to

$$t = 25$$

and

$$CO_2C(25) = 354.0 \text{ ppm}.$$

The corresponding point is (25, 354.0).

Step 2. Set up a (t, CO_2C) coordinate system with t on the horizontal axis, CO_2C on the vertical axis, and plot the four points. (The zigzag at the base of the vertical axis indicates that part of the vertical axis has been left out.) It looks like Figure 3.2.

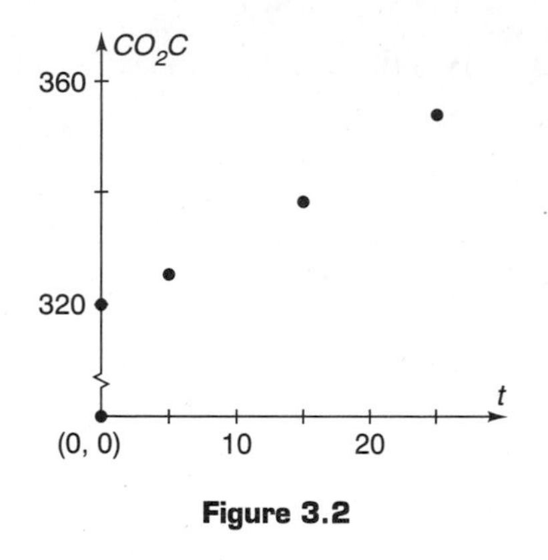

Figure 3.2

Step 3. In order to predict future concentrations of CO_2 based on past information, we need a function whose graph comes close to the points plotted. In this step, we simply look closely at the plot of these points to decide what kind of graph it is shaped like. These points appear to fall very close to a straight line. So, what we have to do is write the equation of a line which comes close to these points. There are many ways to do this, one of which is to choose two of the plotted points and write the equation of the line which passes through them. That's what we're going to do. First we try the years 1965 ($t = 0$) and 1990 ($t = 25$). The corresponding points are (0, 319.9) and (25, 354.0). The slope of the line through these points is

$$m = \frac{354.0 - 319.9}{25 - 0} = 1.36 \text{ (rounded)}.$$

Now use the point-slope formula with point (0, 319.9):

$$CO_2C - 319.9 = 1.36(t - 0),$$

or

$$CO_2C = 1.36t + 319.9.$$

Using functional notation, we write

$$CO_2C(t) = 1.36t + 319.9.$$

Figure 3.3 is the graph of the original data and the "predictor" line whose equation we just wrote.

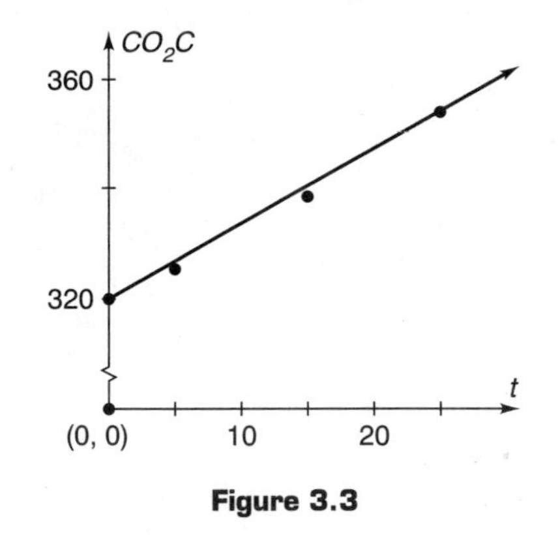

Figure 3.3

If we choose the two points (5, 325.3) and (15, 338.5) (corresponding to which years?), we get the linear function

$$CO_2\,C(t) = 1.32t + 318.7,$$

and here's its graph (Figure 3.4) along with the original data.

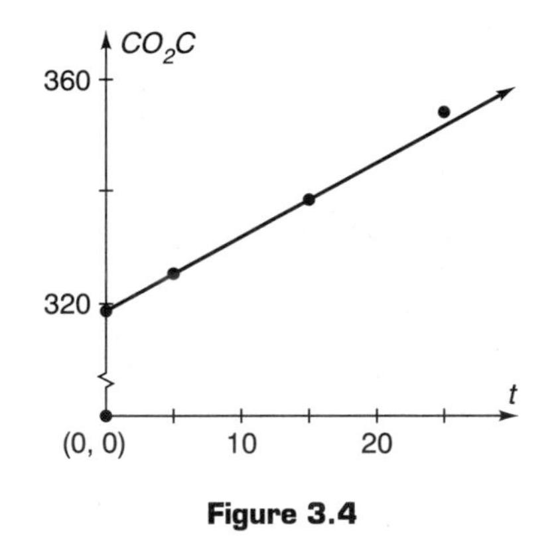

Figure 3.4

We're still on Step 3. There are a total of six linear equations to write by choosing pairs of points, and we've only done two. Be sure to use your calculator for the computations. You may even figure out a program which will do everything for you.

Derive the equations of the remaining lines.

Step 4. By the time we reach this step, we should have equations for all possible lines. We must now decide how to find the "best" line to use to model our data. Remember, we want the one which most closely approximates the real data. This means that the value of the derived function at each of the data years should be pretty close to the CO_2 concentration for that year. We check the first function,

$$CO_2\,C(t) = 1.36t + 319.9.$$

All evaluations in Table 3.2 are done on the calculator.

TABLE 3.2

Year	$CO_2\,C(t)$	Actual CO_2 Concentration
0	319.9	319.9
5	326.7	325.3
15	340.3	338.5
25	353.9	354.0

This first column is time t = years after 1965; the second column is the derived function evaluated at t (e.g., for year 1970, the "predicted" CO_2 concentration is $CO_2\,C(5) = 326.7$ ppm); and the third column is the original data. Next see how far off the "predicted" value is from the actual concentration. In Table 3.3 below, "difference" refers to the distance

TABLE 3.3

Year	Difference
0	0.0
5	1.4
15	1.8
25	0.1

between the predicted value and the actual value. Remember, the distance between two numbers is the absolute value of their difference. We used the data for 1965 and 1990 to derive the function

$$CO_2C(t) = 1.36t + 319.9,$$

so for $t = 0$ and 25, the difference should be zero, and it is for $t = 0$. Note, however, that it is 0.1 for $t = 25$; this is a result of round-off. The differences for years which are used to derive the equation should always be small. (If these differences are large, you may want to check the derivation of your equation for mistakes.)

Next, we need to take into consideration the difference for each of the data years; we'd like all of these differences to be small for the best line, so we define an "overall" *error* for the line to be the sum of these differences. The error for our first derived predictor function is

$$E_1 = 0.0 + 1.4 + 1.8 + 0.1 = 3.3.$$

Of all possible predictor functions for the data provided, we define the best one to be the one with smallest error.

Here's the computation of the error for the second derived predictor function, $CO_2C(t) = 1.32t + 318.7$:

TABLE 3.4

Year	$CO_2C(t)$	Actual CO_2 Concentration	Difference
0	318.7	319.9	1.2
5	325.3	325.3	0.0
15	338.5	338.5	0.0
25	351.7	354.0	2.3

Add the column on the right; if the answer is smaller than $E_1 = 3.3$, then Function 2 is the winner. $E_2 = 3.5$, so Function 1 is better than Function 2.

In summary, the four steps to modeling are:

1. make points out of the data;
2. set up a coordinate system and plot the points;
3. decide on the type of predictor curve to use by observing the shape of the plotted points. Then derive all possible equations. (This is the long step, but without your calculator, you might spend all weekend doing it.)
4. Compute the error for each predictor function to determine the best function. This is also kind of long, but the calculator makes it easy. Be sure to clearly state what the variables mean.

There are four more lines, so we may not have the best one yet. Compute the errors for all possible predictor functions and determine the best one.

Since so much carbon dioxide emission is a result of human activity, a model of population would be interesting. The table below gives U.S. population in millions in the indicated year. Model this with a linear equation using $t = 0$ in 1960. You should go through the four steps listed in this section.

Year	U.S. Population ($\times 10^6$)
1960	180.7
1970	205.1
1980	227.7
1985	238.5
1990	249.9
1992	255.5

Source: *Statistical Abstract of the United States, 1993*, Bureau of the Census.

3.2 USING YOUR FIRST MODEL

First, we list (Table 3.5) all of the linear equations derived from the CO_2 concentration data, together with the error for each. You can use this listing to check your own derivations, and your best equation.

TABLE 3.5

Years	Points	Equation: $CO_2C(t)$ =	Error
1965, 1970	(0, 319.9), (5, 325.3)	$1.08t + 319.9$	9.5
1965, 1980	(0, 319.9), (15, 338.5)	$1.24t + 319.9$	3.9
1965, 1990	(0, 319.9), (25, 354.0)	$1.36t + 319.9$	3.3
1970, 1980	(5, 325.3), (15, 338.5)	$1.32t + 318.7$	3.5
1970, 1990	(5, 325.3), (25, 354.0)	$1.44t + 318.1$	3.1
1980, 1990	(15, 338.5), (25, 354.0)	$1.55t + 315.25$	6.95

The smallest error occurs for years 1970 and 1990, and hence the best function for this model is

$$CO_2C(t) = 1.44t + 318.1.$$

Remember, $CO_2C(t)$ = carbon dioxide concentration in ppm in year $1965 + t$. It is interesting to interpret the slope of this equation: each year the CO_2 concentration will increase by 1.44 ppm.

Here's how this model can be used to predict future atmospheric CO_2 levels.

EXAMPLE 3.1

In the year 2000, $t = 2000 - 1965 = 35$, and the predicted CO_2 concentration is $CO_2C(35) = 368.5$ ppm. ▲

EXAMPLE 3.2

In the year 2050, $t = 2050 - 1965 = 85$, and the predicted CO_2 concentration is $CO_2C(85) = 440.5$ ppm. ▲

EXAMPLE 3.3

The preindustrial level of carbon dioxide concentration was 280 ppm; double this amount is significant, as you will see in the next chapter. Your model can be used to predict when this doubling will occur. Twice 280 is 560 ppm, so we should determine t when $CO_2C(t) = 560$, or solve this equation:

$$\begin{aligned} 560 &= 1.44t + 318.1 \\ 241.9 &= 1.44t \\ t &= \frac{241.9}{1.44} \\ t &= 168 \text{ years.} \end{aligned}$$

This is the year 1965 + 168, or 2133. This is a long time away and may be inaccurate, but the prediction is based on current trends.

The preindustrial carbon dioxide concentration level of 280 ppm actually remained relatively constant until World War II. Let's see how closely our model reflects this information; that is, according to our equation, in what year was $CO_2C = 280$? Thus, we must solve the equation

$$\begin{aligned} 280 &= 1.44t + 318.1 \\ -38.1 &= 1.44t \\ t &= \frac{-38.1}{1.44} \\ t &= -26.5 \text{ years.} \end{aligned}$$

This corresponds to mid 1938 ($1965 - 26.5 = 1938.5$). Not a bad approximation, since the war actually started in 1939!

Notice that this model for CO_2 concentration yields carbon dioxide levels which are below the actual level of 280 ppm for all years before 1938; hence the equation is valid only after 1938. For this reason, we adjust this equation so that $t = 0$ in 1939, and then restrict its domain to $t \geq 0$. This adjustment can be performed by replacing t in $1.44t + 318.1$ by $t - 26.5$ (recall $t = -26.5$ when $CO_2C = 280$). We get

$$\begin{aligned} CO_2C(t - 26.5) &= 1.44(t - 26.5) + 318.1 \\ &= 1.44t + 280. \end{aligned}$$

We still call this last expression $CO_2C(t)$, but now

$$CO_2C(t) = 1.44t + 280,$$

where $CO_2C(t)$ = carbon dioxide concentration in ppm in year

$$1939 + t, \quad t \geq 0. \quad ▲$$

This is the model we use for the remainder of Part I.

Use this revised function for CO_2C to answer the following questions.

1. Predict CO_2 concentration in the year 2000.
2. In what year will the concentration be double the preindustrial level? (For Questions 1 and 2, you should get the same answers as the ones in the last section using the old equation.)
3. Predict CO_2 concentration in year 2061. (This is the year Halley's Comet returns.)
4. In what year will CO_2 concentration be 420 ppm?
5. Predict when the air will be all CO_2. Do you think this is a reasonable use of the model?
6. Discuss what social, political, or physical changes might affect the accuracy of the CO_2 concentration model.

Use your U.S. population model (see data at end of Section 3.1) to answer the following questions.

1. Predict the U.S. population in 2020.
2. When will the U.S. population double its present population?

3. Write a verbal interpretation of the slope of your population equation, that is, how much does the U.S. population increase annually?
4. Predict the U.S. population in the year you were born. You might want to go to the library and check the accuracy of this prediction in the *Statistical Abstracts*.
5. Currently, the U.S. population is 4.7 percent of world population. Assuming that this percentage remains constant, predict the year in which world population will reach 8 billion.
6. Discuss any social, political, or other factors which might affect the accuracy of the population model.

In the next chapters, you'll discover some consequences of more and more CO_2 in the atmosphere.

Composite Functions

4.1 DEFINITION AND EXAMPLES

Finding a composite function is as easy as evaluating a function. Actually, it's almost the same as the operation of evaluation of a function at a number, except you substitute another function for x instead of a number. Just remember the cat ( ), Section 1.1: whatever goes in the parentheses for x also replaces x in the mathematical expression.

If $f(x)$ and $g(x)$ are two functions, then $f(g(x))$ denotes the composite function obtained by substituting the function $g(x)$ in for the x in the function $f(x)$. This is read "f of g of x." You may be wondering why anyone would ever need to find a composite. Often it is desirable to express mathematically a relation between two entities. Remember the earlier function $F(w)$, in Section 1.1, which defines the number F of catfish in a north Georgia lake in terms of the number w of gallons of toxic waste dumped upstream. Well, there's also a restaurant on that lake which serves fresh catfish, and the price of a dinner depends on the availability of fish from the lake. So think about the function $P(F)$ which defines the price P of a catfish dinner in terms of the number of fish in the

Note: Chapter 4 is a prerequisite for Chapter 5.

lake. (Note: this would be a decreasing function surely; the fewer the fish, the higher the price.) The real cause of a price increase is the amount of toxic waste dumped. So, the function which would describe this is the composite $P(F(w))$; this defines the price of a catfish dinner in terms of the amount of toxic waste dumped upstream! See Exercise 9 at the end of this section for more on toxic waste and catfish dinners.

We provide some examples of finding composites using simple algebraic functions.

EXAMPLE 4.1

Let $f(x) = x^2 + 4x + 3$, and $g(x) = 2x$. To find $f(g(x))$ substitute $g(x)$ (which equals $2x$) for x in $f(x)$ to get

$$f(g(x)) = f(2x) = (2x)^2 + 4(2x) + 3 = 4x^2 + 8x + 3.$$

Also, $g(f(x))$ is a composite. This time you substitute $f(x)$ (which equals $x^2 + 4x + 3$) for x in $g(x)$ to get

$$\begin{aligned} g(f(x)) &= g(x^2 + 4x + 3) = 2(x^2 + 4x + 3) \\ &= 2x^2 + 8x + 6. \quad \blacktriangle \end{aligned}$$

EXAMPLE 4.2

Let $f(x) = 3x - 1$ and $g(x) = x^2 + 7$.

$$f(g(x)) = f(x^2 + 7) = 3(x^2 + 7) - 1 = 3x^2 + 20$$

and

$$\begin{aligned} g(f(x)) &= g(3x - 1) = (3x - 1)^2 + 7 = 9x^2 - 6x + 1 + 7 \\ &= 9x^2 - 6x + 8. \quad \blacktriangle \end{aligned}$$

One thing you should note: a composite of two functions yields another function. There is no equation to solve or anything like that. (Students for thousands of years have found a composite, then set it equal to zero and tried to solve for x; who knows why?)

EXAMPLE 4.3

Let $R(a) = \sqrt{a^5 + 1}$ and $S(b) = b^2$.

$$R(S(b)) = \sqrt{(b^2)^5 + 1} = \sqrt{b^{10} + 1}. \quad \blacktriangle$$

EXAMPLE 4.4

Let $T(c) = c^2 - 2.4c + 6.8$ and $U(d) = 3d + 4.7$, and determine $T(U(0))$ and $U(T(1))$.

Here you could first determine the composites $T(U(d))$ and $U(T(c))$ and then evaluate these at 0 and 1, respectively. However, perhaps the easier way is to first determine $U(0)$ and then evaluate $T(c)$ at this number; similarly for $U(T(1))$.

$$U(0) = 3(0) + 4.7 = 4.7,$$

and

$$T(U(0)) = T(4.7) = (4.7)^2 - 2.4(4.7) + 6.8 = 17.61.$$

Similarly,

$$T(1) = 1^2 - 2.4(1) + 6.8 = 5.4$$

and then

$$U(T(1)) = U(5.4) - 3(5.4) + 4.7 = 20.9.$$

Of course, all these evaluations can be performed easily with your calculator. ▲

4.1 THINGS TO DO

1. $f(x) = 3x - 5$; $g(x) = 2 - x$. Determine $f(g(x))$.
2. $f(x) = 5(2 - 3x)$; $H(x) = 3 + (2x - 5)$. Determine $H(f(x))$ and $H(f(.2))$.
3. $r(x) = \frac{1}{x}$; $s(x) = 5x + 2$. Determine $s(r(x))$.

4. $AB(t) = t + 2$; $C(s) = s^2 - 2$. Determine $C(AB(-2))$.

5. $E(x) = x$; $H(x) = .13578x^3 - .00023$. Determine $H(E(x))$ and $H(E(1))$.

6. $PQ(x) = 3x + 1.5$; $QP(x) = \frac{x - 1.5}{3}$. Determine $PQ(QP(x))$.

7. $A(r) = \sqrt{r^3 + 5}$; $H(s) = s^2$. Determine $HA(r)$ and $HA(0)$.

8. $A(t) = t$; $R(u) = u^2 + 1$; $T(v) = \sqrt{v}$. Determine $ART(v)$, $RAT(v)$, and $TAR(u)$.

9. A chemical factory has set up business on a certain north Georgia river. This river empties into a large lake, on which is a restaurant which serves fresh catfish dinners. The chemical factory started dumping toxic waste into the river, which reduced the catfish population. This is described by the equation

$$F(w) = 50 - .5w,$$

where F = number of fish per acre and w = gallons of toxic waste dumped per day. In turn, the reduced availability of catfish caused the restaurant to increase the price of its dinner. This is described by

$$P(F) = 24 - .4F, \text{ where } P = \text{price of a catfish dinner.}$$

a. Determine the composite function which defines the price of a catfish dinner in terms of the number w of gallons of toxic waste dumped by the chemical factory.
b. How much did a dinner cost before the factory opened?
c. How much waste will correspond to a price of $12?
d. How much waste will kill all the fish?

The next problem is dedicated to Dine' CARE (Citizens Against Ruining the Environment).

10. A deadly virus is carried by a certain kind of mouse which dwells in and about the floor of a 200 acre old growth forest in the southwestern United States. On the edge of this forest is a small village with a population of 23. When the mouse population gets too large, the mice venture more into the village in search of food and new places to inhabit, thereby spreading their virus and causing villagers to become infected. There is also a certain kind of owl which dwells in this forest, and as we all know, owls love to feast on mice. Since the virus is only harmful to humans, the owls are unaffected so that as long as the number of owls stays at a reasonable level, the mice population is kept at a comfortable number and the virus is not transmitted to the villagers. However, visualizing large profits, the owner of the old growth forest begins to sell timber to a large lumber company; this, of course, removes the homes of the owls, and they either die off or must leave in search of other shelter.

The interrelationships of the villagers, virus, mice, owls, trees, and loggers can be described by the following mathematical model.

$$\begin{aligned} V &= \text{number of cases of virus among the villagers,} \\ M &= \text{mouse population,} \\ W &= \text{owl population,} \\ T &= \text{number of acres of trees remaining.} \\ V &= .025M - 20 \\ M &= 1600 - 20W \\ W &= .2T \end{aligned}$$

Use this model to answer these questions.

a. What are the domain and range of each of the three functions in the model?

b. What is the maximum number of owls which can survive in the forest? What is the maximum number of mice which can survive? What is the maximum number of viral cases which could occur among villagers?

c. Determine the composite which describes the number of mice in terms of the number of acres of timber remaining in the forest. Also determine the composite which describes the number of cases of viral infection among the villagers in terms of the number of acres of trees remaining.

d. How many cases of virus will occur if the forest is untouched? if half the forest is logged? if three-fourths is logged?

e. How many villagers will be left if the entire forest is destroyed?

CHAPTER 5

Global Temperature and Ocean Level

5.1 ON THE BEACH . . .

If the global temperature rises, then water expands, and polar ice caps would melt, thus causing ocean levels to rise. Should this happen, our beaches will disappear.

It is estimated that a doubling of the preindustrial level of atmospheric CO_2 concentration will cause an average global temperature increase of 2° F to 8° F, and furthermore, an increase of as little as 3° F in global temperature can cause a one foot rise in ocean levels. That's pretty significant!

Figure 5.1 shows long term data for atmospheric carbon dioxide concentration and average global temperature.

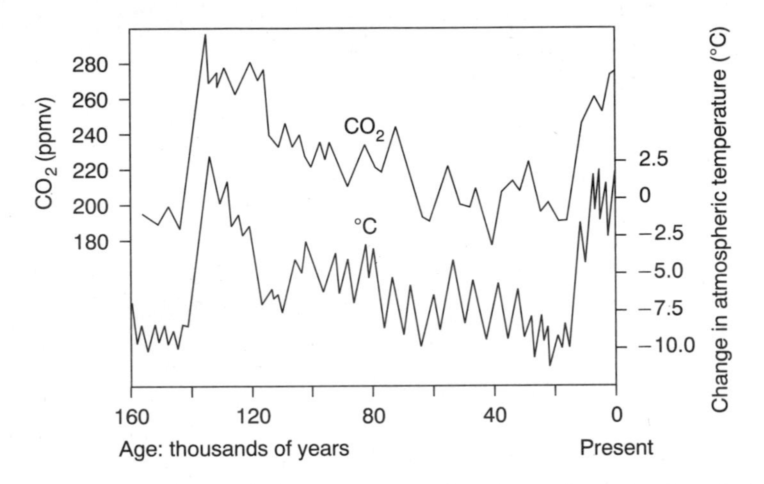

Figure 5.1 The Vostok record of temperature and concentrations of carbon dioxide in the atmosphere, versus time (Figure, "The Vostok record of temperature and concentrations of carbon dioxide in the atmosphere, versus time" from *New Scientist,* April 14, 1988, page 28. Reprinted by permission of *New Scientist,* London.)

What we are going to do now is to study what could actually happen to the beaches if CO_2 emission continues to increase at its current rate. First, we write an equation which relates temperature increase to atmospheric CO_2 concentration. We derive the equation using the estimate that doubling the preindustrial level of CO_2 causes a 4.5° F temperature increase, and assume the relationship is linear. (The "average" estimate is 5° F, and for our assumption, we take an estimate a little below this average.) We use a (CO_2C, GT) coordinate system, where the variables

$$CO_2C = \text{carbon dioxide concentration in ppm,}$$

and

$$GT = \text{global temperature increase since 1939;}$$

that is, $GT = 0°$ F in 1939. (Remember the significance of the year 1939? According to the CO_2 concentration equation from Chapter 3, the carbon dioxide concentration was at the preindustrial level of 280 ppm in 1939.) To write our equation, we need two points. The first one comes from the preindustrial level CO_2 concentration, which was 280 ppm, and the corresponding global temperature increase is 0° F; so the point is (280, 0). To get the second point, use the information above: double 280 ppm is 560 ppm, and the corresponding $GT = 4.5°$ F, so this point is (560, 4.5). The slope of the line is

$$m = \frac{4.5 - 0}{560 - 280} = .016,$$

and the equation is

$$GT - 0 = .016(CO_2C - 280)$$

or $$GT = .016CO_2C - 4.48.$$

(Here we round the slope to three places for more accuracy.) In functional notation,

$$GT(CO_2C) = .016CO_2C - 4.48,$$

where

$$GT(CO_2C) = \text{global temperature increase,}$$
$$CO_2C = \text{carbon dioxide concentration in ppm.}$$

The significance of the slope of the equation is this: an increase in CO_2 concentration of one ppm corresponds to an increase in global temperature of .016° F.

Next, we can relate global temperature increase to year by using our $CO_2C(t)$ model derived in Chapter 3:

$$CO_2C(t) = 1.44t + 280,$$

where

$$CO_2C(t) = \text{carbon dioxide concentration in ppm in year } 1939 + t.$$

We define this relationship between GT and t by finding the composite function $GT(CO_2\,C(t))$.

$$GT(CO_2\,C(t)) = .016(1.44t + 280) - 4.48$$

or

$$GT(CO_2\,C(t)) = .023t.$$

This composite function defines global temperature increase GT in terms of year t, where $t = 0$ in 1939. Notice that the slope of this linear function tells you that the average global temperature will increase .023° F each year.

Write a linear equation which defines the increase in ocean level in terms of the increase in global temperature. Use a (GT, OL) coordinate system, where OL = ocean level increase since 1939; this defines a function $OL(GT)$. To derive this equation, use the information that a temperature increase of 3° F corresponds to a one foot rise in ocean level. Also, note that in 1939 temperature increase is 0° F, and oceans have not risen any. This will give you two points.

After deriving $OL(GT)$, form the composite $OL(GT(CO_2\,C))$ which defines ocean level increase in terms of carbon dioxide concentration.

Next, obtain the important composite function which defines ocean level increase in terms of year t. To derive this function, find your $CO_2\,C(t)$ function where $t = 0$ in 1939, and form the composite $OL(GT(CO_2\,C(t)))$. Using this linear function determine how much the oceans will rise each year.

This last function, $OL(GT(CO_2\,C(t)))$, is quite valuable; we simplify its name to $O(t)$. It gives us a tool for figuring how many years it will take for the ocean to rise a given distance. For example, if it is important to know when the ocean will have risen three and one-half feet above its 1939 level, then simply take $O(t) = 3.5$, solve the equation for t, and add the answer to 1939.

EARTH NOTE Migrating Islands

The barrier islands off the coast of the middle Atlantic are moving inland. Geologists believe that these islands were developed before the end of the last ice age (about 13,000 years ago), a time when the sea level was about 350 feet lower and the coast line about 20 to 75 miles further east. Storms and high waves move the sand from the ocean side beaches to the bay side, causing the islands to "creep" inland. Beach erosion ranges from 2–5 feet per year in Maryland and North Carolina to 10–25 feet per year in Virginia. However, the rise in ocean level in this region during the last century, about one foot, creates deeper channels between the islands and the mainland. These deeper channels contribute to more waves, which causes erosion of the mainland shore. Thus, the net result is that the shoreline is moving westward at an even faster rate than the barrier islands. (*Washington Post National Weekly Edition*, Sept. 13–19, 1993)

PROBLEM: An island leaves North Carolina in 1993 and heads north at the rate of 4 feet per year. When the island reaches Virginia, it increases its rate of travel to 21 feet per year and heads due west. When will the island reach Chicago?

5.2 OR, WHAT BEACH?

On to the beach! In 1991, the authors drove to Tybee Island (near Savannah, Georgia) and estimated the average height of the sea wall on the beach there to be four feet, and the base of this wall to be 2.2 feet above the ocean level. See the figure on the next page.

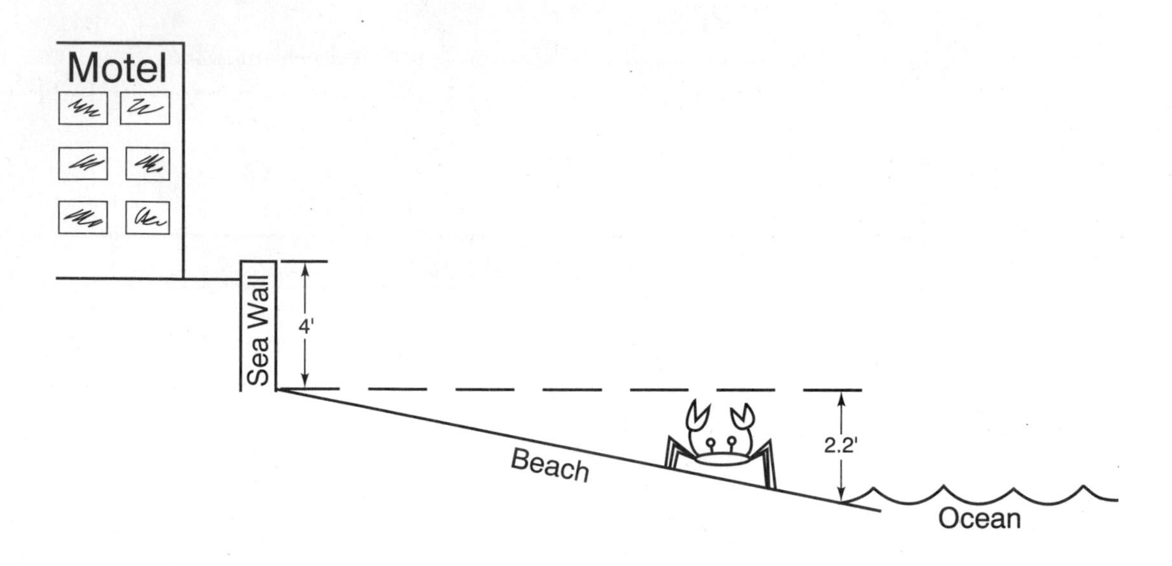

EARTH NOTE High Rises and Weather Patterns

Did you know that tall buildings affect weather patterns? Some years ago, thunderstorms and rain clouds which developed over the Gulf of Mexico would move inland over the west coast of Florida, proceed across the state and out over the Atlantic providing ample rainfall for inland areas and the eastern coast. However, in the past twenty years or so, there has been extensive development of high rise hotels and condominiums along the eastern shore. These large buildings create thermals and cause weather systems to turn north as they come inland so that they never reach the eastern and central areas of Florida. The result has been periodic drought conditions in some portions of the state. In particular, the Everglades have suffered considerably from these conditions, as well as from other human development projects such as damming and irrigation methods upstate.

Find out what year the ocean will completely cover the beach at Tybee Island. Next, it would also be interesting to know when the ocean will be over the sea wall and flood the development. When will this happen?

EARTH NOTE Ocean Level Changes

The actual ocean level change depends on the ocean, the latitude, and other geographical characteristics. On the Atlantic, the trend in changing sea level ranged from an increase of 0.6 millimeters per year in Boston to 3.2 mm per year in Charleston, South Carolina, while the Pacific ranged from a low increase of 0.7 millimeters per year in San Diego to a high of 3.2 millimeters per year in Seattle. How do these figures compare with the estimates of sea level change predicted by your model? (Watch your units.)

PART II

Factors Contributing to Carbon Dioxide Build-up

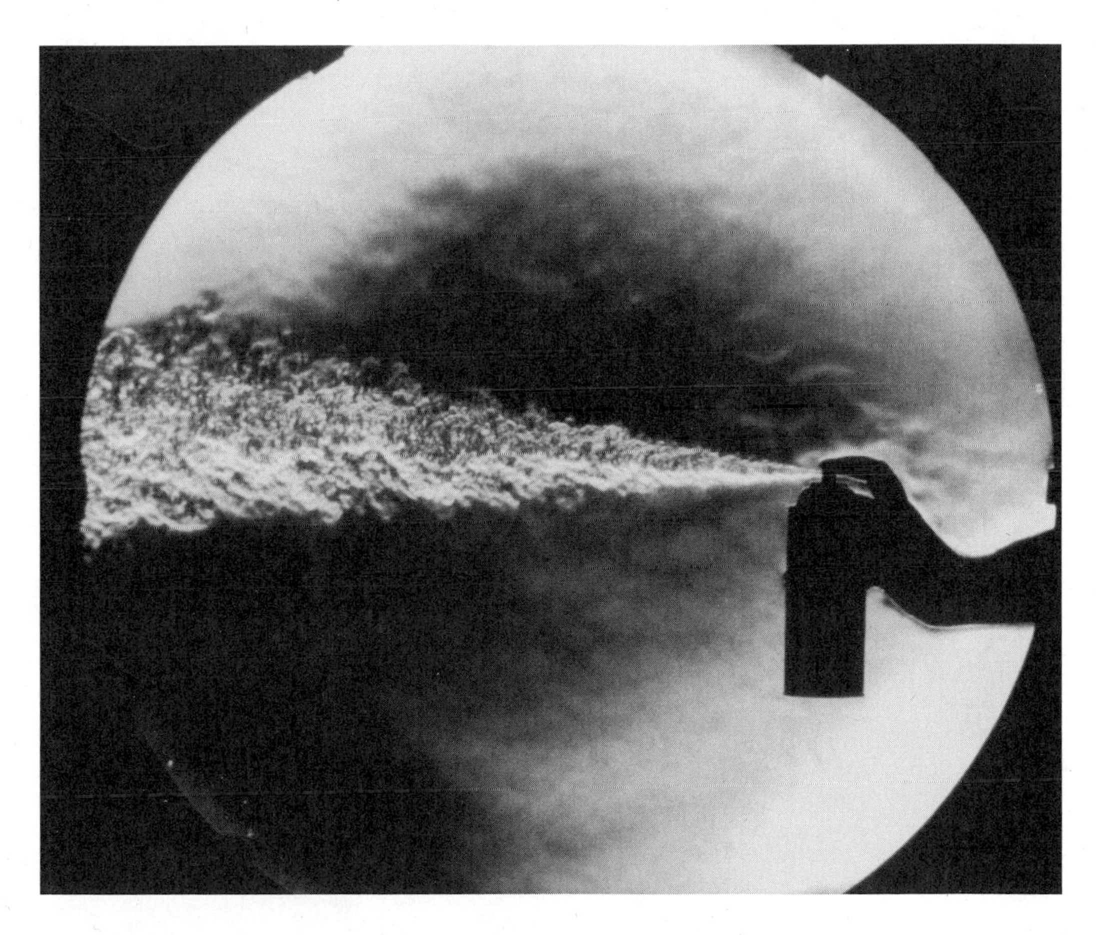

Introduction

In Part II you will study three major sources of carbon dioxide emission: automobiles in the United States, energy consumption in the United States, and deforestation.

If you live in the suburbs near a big city, and work a nine-to-five job in the city, then you probably already know about carbon dioxide emission from automobiles. If you do not drive to work and home every day in that sea of metal, through that fog of carbon compounds, then may you never have to.

Cars are one of the major sources of carbon dioxide in the United States. The number of automobiles continues to increase sharply, and Americans are driving more and more each year. Although most manufacturers are improving the fuel efficiency of their vehicles, the resulting improvements are not fast enough to balance CO_2 emissions. It was estimated that in the year 1991, over 750 million tons of carbon dioxide were pumped into the atmosphere from passenger vehicles in this country. This figure represents an 18 percent increase over the corresponding emission ten years ago. Don't breathe too deeply.

The United States is responsible for 25 percent of the global emission of carbon dioxide, and 20 percent of that is due to automobiles. Thus, automobiles in the U.S. contribute 5 percent of the world's CO_2 emission.

Chapters 6, 7, and 10 cover additional mathematical topics needed for Part II. In Chapter 8 we study the output of carbon dioxide from automobiles in the United States. Three sets of data are easily available from the *Statistical Abstract of the United States:* number of cars by year; the average number of miles each car travels per year; and the average gas mileage per car by year. (Copies of the *Statistical Abstract* are kept in most libraries, and are updated each year.) These three pieces of information, together with the fact that each gallon of gasoline burned results in approximately 20 pounds of CO_2 can be combined to determine the annual carbon dioxide emission.

The gasoline you burn in your car is just one source of carbon dioxide emission. Every time you turn on a light bulb or run your furnace, you use energy. It takes energy to manufacture the paper that this is written on and to grow the food you eat. The production and consumption of energy is one of the major causes of greenhouse gas emissions. In the United States, the annual average consumption of energy is equivalent to the consumption of 2200 gallons of petroleum per person. For comparison, the average person in Japan uses the equivalent of about 1000 gallons of petroleum per person, and in the developing countries such as Brazil, China, or India, the consumption is the equivalent of 150 gallons per capita. Energy consumption dropped in the early 1980s but since then it has been on the rise.

In Chapter 9 energy consumption data in the United States are provided (*Statistical Abstract of the United States*) for coal, petroleum, and natural gas. Each set of data is to be modeled by the groups, and finally total energy consumption will be determined.

Another major factor affecting global warming is the alarming rate of destruction of tropical rain forests. Almost all of this destruction, or deforestation, is done for human consumption.

Many acres of rain forest are being cleared and burned for homesites and farming by locals, and then more land needs to be cleared to build access roads, and on and on. It took years upon hundreds of years for rain forests to form, and once destroyed they don't grow back. When the trees are cut and burned, they release tons of carbon into the atmosphere, thus adding to the already increasing volume of greenhouse gases. Trees also absorb carbon when growing, so when forests are cut, much of the carbon which would have been absorbed remains in the atmosphere. Tropical forests can absorb over 80 metric tons of carbon per hectare per year, and when cut release up to one-half metric ton of carbon for each hectare. Therefore, deforestation has two negative effects on a stable planetary climate.

And besides all this, rain forests are very beautiful. And so are the animals that make their homes there.

In Chapter 11 we categorize most of the direct reasons for deforestation into three big areas: (1) logging; (2) cattle grazing; and (3) agriculture, mining, and development. Indirectly, these factors are related to population growth and increasing demand for forest products. We analyze each of these areas to find out its effect on the overall loss of rain forests and on atmospheric carbon dioxide.

In Chapter 12, we build models which define total emission of carbon from each of the three sources, automobiles, energy consumption, and deforestation.

The mathematical concepts used to conduct these studies in Part II are linear, quadratic, logarithmic and piecewise functions, systems of equations, and matrices. Also, exponential functions are introduced for use later.

A Reminder. Coefficients of functions and equations in all environmental chapters should be rounded as follows: linear, two places; quadratic, exponential, and logarithmic, four places. Sometimes we deviate from this practice but not without warning.

Some discrepancy may occur in answers if round-off is done during calculation as opposed to only rounding the final result. In all situations the final answer should be rounded to the same number of places as the original data.

Quadratic Functions

6.1 NOTATION AND DEFINITIONS

You are moving up a notch! In the preceding section, you studied linear functions. The largest exponent used for a linear function is 1; quadratic functions have x raised to the second power: x^2. A *quadratic function* looks like

$$f(x) = ax^2 + bx + c,$$

where a, b, and c are just constants, but important ones. The equation above is referred to as the *standard form* for a quadratic function. The graph of a quadratic function is called a *parabola* (pa rab' o la, not pa ra bo' la), and it looks like one of the two curves in Figure 6.1.

You get the curve on the left when the leading coefficient a is positive, and people say that this parabola *opens up.* On the other hand, if $a < 0$, the parabola on the right occurs, and then people say that it *opens down.*

Note: Chapter 6 is a prerequisite for Chapters 8 and 9.

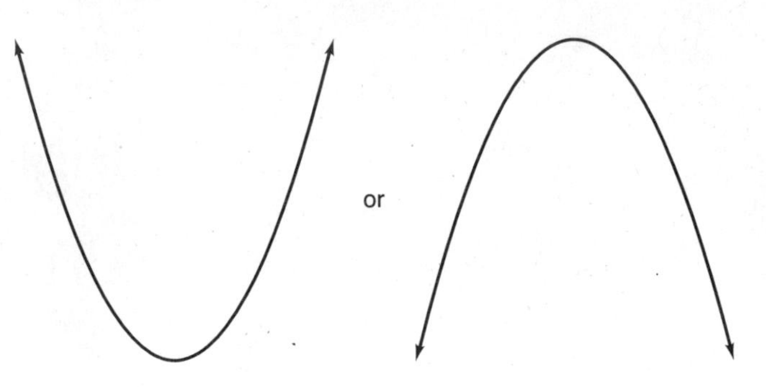

Figure 6.1

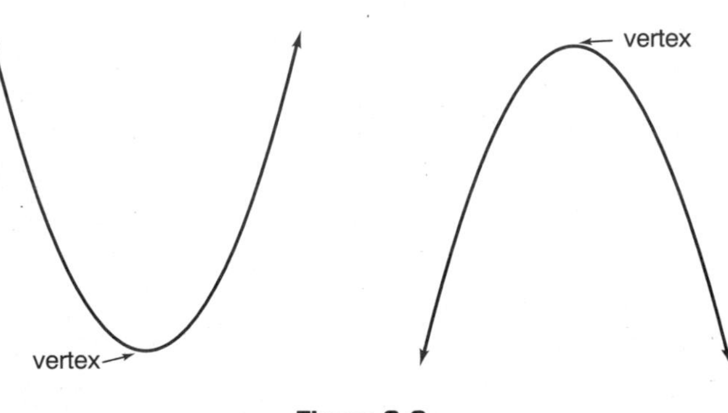

Figure 6.2

Every parabola in the universe has a *vertex*, which is quite important. The vertex is the lowest point ($a > 0$), or the highest point ($a < 0$) on the graph of a quadratic function. See Figure 6.2.

If you draw a vertical line through the vertex of a parabola, its graph on one side of the line is the mirror image of its graph on the other side; that is, it is symmetric about that vertical line.

Other points on a parabola that are really important are its intercepts. A parabola can have as many as three intercepts, one on the y-axis, and possibly two on the x-axis (recall that intercepts are points where a graph crosses an axis).

The domain of a quadratic function consists of all real numbers x, and the range will be discussed later when we talk about the vertex in more detail.

Right now, we concentrate on the important features of the graph of the quadratic function $y = ax^2 + bx + c$:

1. open up or down?
2. y-intercept?
3. x-intercept(s)?
4. vertex?

and

5. maximum or minimum value?

The important thing about those constants a, b, and c is that you can draw the graph and determine those five items by just doing some arithmetic with a, b, and c, or letting your calculator do it for you. We'll show you how, step-by-step.

Step 1. Open up or down? This is simple. Put your function in standard form, and look at a. If $a > 0$, it opens up. If $a < 0$, it opens down.

In the following sections, we go through the remaining items with you, so you can be an expert parabola artist.

6.2 INTERCEPTS

Step 2. y-intercept? Easy. Substitute zero for x in the equation $y = ax^2 + bx + c$ to get the answer. Remember, you're looking for the place where the graph crosses the y-axis, which is where $x = 0$.

Step 3. x-intercept, or x-intercepts, or, are there any? Now you're looking for the places where the graph intersects the x-axis, and this happens where $y = 0$. So, you substitute $y = 0$, and then you have to solve the quadratic equation

$$0 = ax^2 + bx + c.$$

There are three possibilities which can occur. We look at each graphically for a parabola which opens up. See Figure 6.3.

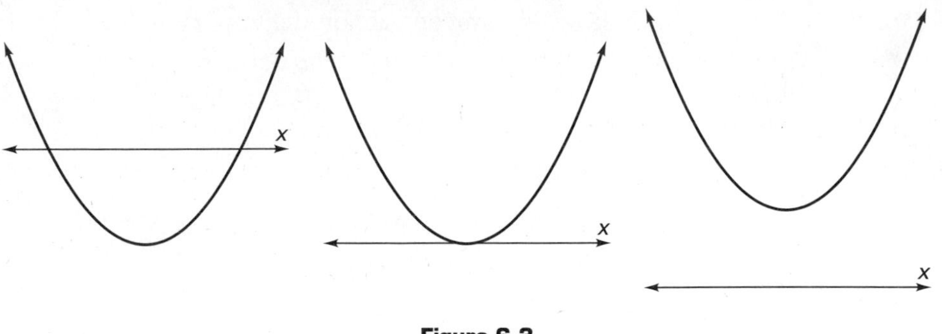

Figure 6.3

Of course the same three situations can happen for a parabola which opens down—turn the page over and upside down, hold it up against a window and look at these graphs.

When you get two x-intercepts, there will be two solutions to

$$0 = ax^2 + bx + c;$$

when there is one x-intercept, there'll be one solution; and when there is no x-intercept, the equation has no solution.

We review ways to solve quadratic equations. Probably factoring is what comes immediately to your mind. For example: solve

$$0 = x^2 - x - 2.$$

Factor $0 = (x - 2)(x + 1)$, and set each factor equal to zero.

$$x - 2 = 0, \qquad x + 1 = 0;$$
$$x = 2 \quad \text{or} \quad x = -1.$$

However, there's not one single quadratic equation in *Earth Algebra* that can be easily factored.

But, remember the quadratic formula? (See Appendix C for its derivation.) This can be used to solve any quadratic equation, if it has a solution. It always works. *Always.* Just put your equation in standard form, pick off the coefficients a, b, and c and plug them into

$$x = \frac{-b \pm \sqrt{b^2 - 4ac}}{2a}.$$

This is the celebrated *quadratic formula;* if you don't know it, or have forgotten it, learn it now. You'll need it throughout *Earth Algebra.*

There's one thing you must remember always to do before solving a quadratic equation—no matter how it presents itself, you must put it in standard form before using the quadratic formula. Here are some examples.

EXAMPLE 6.1

Find the x-intercepts of the quadratic function

$$y = -9.6 - 3.2x + 1.5x^2.$$

Set $y = 0$,

$$0 = -9.6 - 3.2x + 1.5x^2,$$

and solve.
Use the quadratic formula, but first rearrange to get

$$0 = 1.5x^2 - 3.2x - 9.6.$$

Now $a = 1.5$, $b = -3.2$, $c = -9.6$, and

$$x = \frac{-(-3.2) \pm \sqrt{(-3.2)^2 - 4(1.5)(-9.6)}}{2(1.5)}$$

Get your calculator! Your answers are (rounded to two places) $x = 3.81$ or $x = -1.68$. Graphically, the quadratic $f(x) = 1.5x^2 - 3.2x - 9.6$ has two x-intercepts. See Figure 6.4. ▲

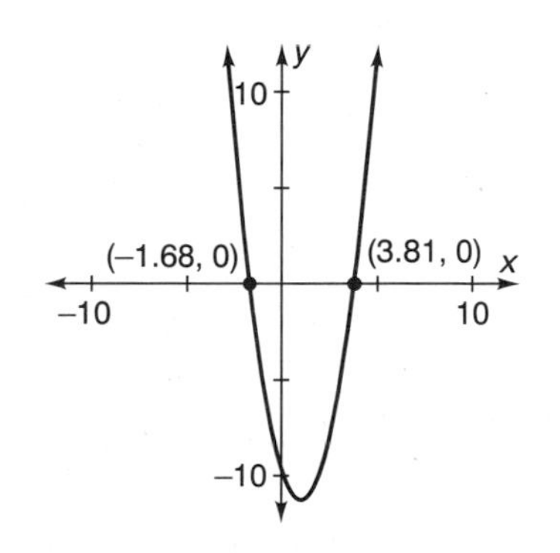

Figure 6.4

EXAMPLE 6.2

Determine the x-intercepts of the quadratic function

$$f(x) = -2.3x^2 + 1.1x - 27.6.$$

Replace $f(x)$ by 0 and solve the quadratic equation

$$0 = -2.3x^2 + 1.1x - 27.6.$$

The equation is already in standard form, and $a = -2.3$, $b = 1.1$, $c = -27.6$.

$$x = \frac{-1.1 \pm \sqrt{(1.1)^2 - 4(-2.3)(-27.6)}}{2(-2.3)}$$

And, the calculator does your arithmetic and tells you that you have erred! Your error is in trying to take the square root of -252.71. No one can take the square root of a negative number and get a real number for an answer. This means that your original quadratic equation has no real solution. (In *Earth Algebra,* the only numbers we use are real.) Graphically, as shown in Figure 6.5, this happens when there are no x-intercepts. ▲

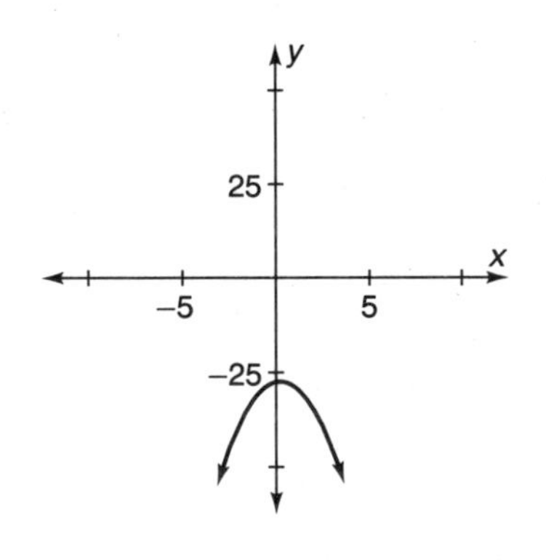

Figure 6.5

EXAMPLE 6.3

Determine the x-intercepts for this quadratic function:

$$y = (x^2 - 9.2x + 26.01) - x.$$

Standard form first:

$$0 = x^2 - 10.2x + 26.01.$$

Now $a = 1$, $b = -10.2$, $c = 26.01$, and

$$x = \frac{-(10.2) \pm \sqrt{(-10.2)^2 - 4(1)(26.01)}}{2(1)}$$

The calculator tells you that the quantity under the radical is zero, so the $\pm$ doesn't matter. There's only one answer: $x = 5.1$. Graphically, this means that the parabola has (5.1, 0) as its only x-intercept. See Figure 6.6. ▲

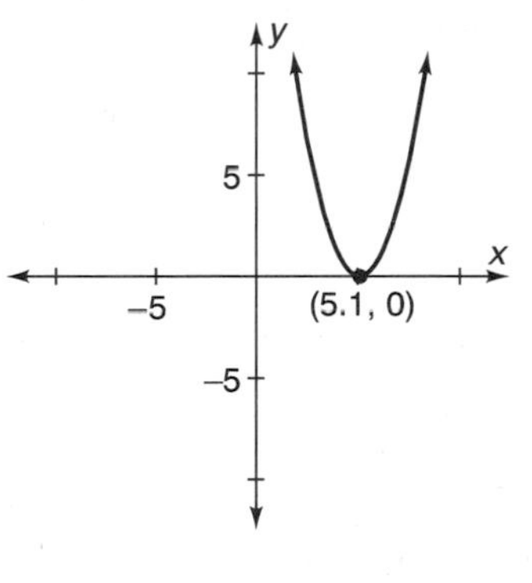

Figure 6.6

6.2 THINGS TO DO

Decide whether each of these parabolas opens upward or downward and determine the coordinates of the x- and y-intercepts of each.

1. $y = x^2 - 6x + 5$
2. $y = 5x^2 + 6x + 15$
3. $y = .02x^2 - 63.1x + 27$
4. $y = 3200t^2 - 450t + 12$
5. $E(z) = -3z^2 + 27.5z + 13.1$
6. $F(x) = .0025x^2 - 6x + 3600$

7. $G(x) = 9 - 4x^2 + .6x$
8. $R(x) = .0027x^2 + 1.2x - .13$
9. $A(x) = x - 3.5x^2$
10. $B(x) = 8x^2 + 2.2$
11. $C(x) = 7.14x^2 + 3$
12. $D(x) = 6x^2 - 54$
13. $f(x) = x^2$. Also, evaluate this function at $x = 111,111,111$. (You may not be able to use your calculator.)

6.3 VERTEX

Step 4. Coordinates of the vertex? The first two coefficients, a and b, tell you what the first coordinate of the vertex is. The formula is:

$$x = \frac{-b}{2a}.$$

You get the second, or y-coordinate by substituting $\frac{-b}{2a}$ into the function; that is, $y = f\left(\frac{-b}{2a}\right)$. The coordinates of the vertex are $\left(\frac{-b}{2a}, f\left(\frac{-b}{2a}\right)\right)$. Compute $\frac{-b}{2a}$ on the calculator, then evaluate the quadratic function at this number.

EXAMPLE 6.4

Find the vertex of the parabola $f(x) = -2.6x^2 + 7.6x - 10$.

Here $a = -2.6$, $b = 7.6$, so the x-coordinate of the vertex is

$$x = \frac{-7.6}{2(-2.6)} = 1.46 \text{ (rounded)}.$$

Next $y = f(1.46) = -4.45$ (also rounded) is its y-coordinate. The vertex is the point $(1.46, -4.45)$. In this case, it is the highest point on the graph. Note that the graph opens down because $a = -2.6 < 0$. See Figure 6.7. (If $a > 0$, the parabola opens up, and the vertex will be the lowest point on the graph.) ▲

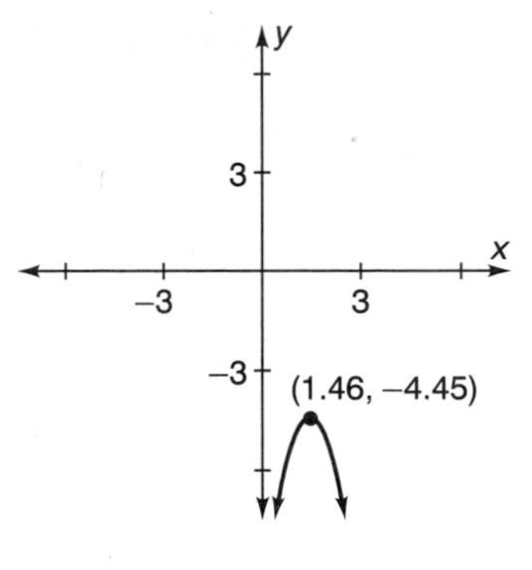

Figure 6.7

EXAMPLE 6.5

Find the vertex of the parabola $y = 2x^2 - 4.6x + 5.1$. Here $a = 2$, $b = -4.6$, so the x-coordinate of the vertex is

$$x = \frac{(-4.6)}{2(2)} = 1.15,$$

and the y-coordinate is obtained by substituting $x = 1.15$ into thc original equation,

$$y = 2(1.15)^2 - 4.6(1.15) + 5.1 = 2.46.$$

Hence the vertex is (1.15, 2.46), and is the lowest point on the graph since this parabola opens up. (See Figure 6.8.) ▲

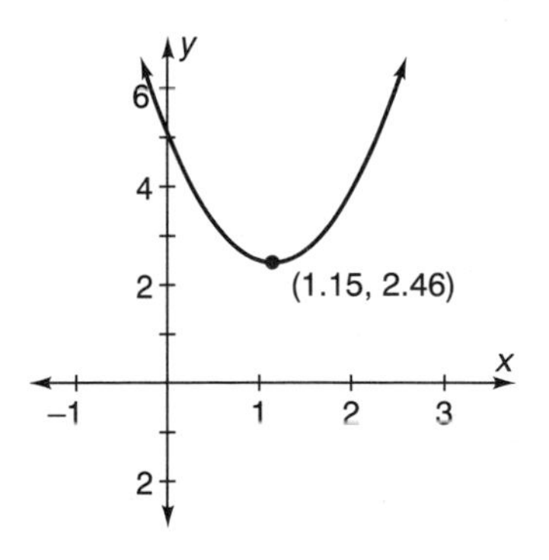

Figure 6.8

EXAMPLE 6.6

Find the vertex of the parabola $f(x) = 8x^2 - 1$. In this case $a = 8$, $b = 0$, so

$$x = -\frac{0}{2(8)} = 0,$$

and $y = f(0) = 8(0)^2 - 1 = -1$, and the vertex is $(0, -1)$. The vertex for this parabola is also the lowest point on the graph. (See Figure 6.9.) ▲

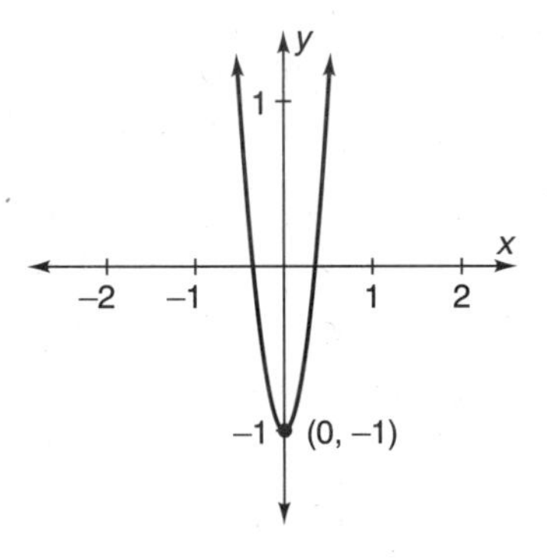

Figure 6.9

6.3 THINGS TO DO

Find the vertex of each parabola.

1. $y = x^2 + 6.2x + 15.7$
2. $y = 3x^2 - 8.1x - 13$
3. $y = 97.1 - 21.4x + 33x^2$
4. $y = x - 3 - x^2$
5. $f(x) = 4x^2 + 6x + 8$

6. $f(x) = 6x^2 + 1$

7. $f(x) = 352\ .1x^2$

8. $f(x) = 8.1 - 4.3x - 5.6x^2$

9. $A(x) = 4x^2 - x$

10. $B(x) = 5 - 2x^2$

6.4 MAXIMUM, MINIMUM, AND RANGE OF A QUADRATIC FUNCTION

Step 5. Maximum or minimum value? The y-coordinate of any point on a graph tells how high (or low) the point is. Since the second coordinate of a point on the graph of a function is $f(x)$, then the highest point on the graph has the largest possible y-coordinate, or functional value, that can occur. So a quadratic function has a maximum when its graph opens down ($a < 0$), and that maximum is the y-coordinate of its vertex. Similarly, if a parabola opens up, it has a minimum which is the y-coordinate of its vertex.

In Example 6.4 above, $f(x) = -2.6x^2 + 7.6x - 10$ has a maximum value of -4.45, which occurs when $x = 1.46$. See Figure 6.7. The quadratic functions in examples 6.5 and 6.6 have minimum values of 2.46 and -1, respectively.

Maximum and minimum values of functions are really important throughout *Earth Algebra.* If you know the maximum or the minimum of a function, you can often determine its range. The function in Example 6.4,

$$f(x) = -2.6x^2 + 7.6x - 10,$$

has as its maximum value -4.45, and has no minimum value, so its range consists of all $y \leq -4.45$. The range of the quadratic function in Example 6.5 consists of all $y \geq 2.46$. It has a minimum of $y = 2.46$ and no maximum. Similarly, the quadratic function in Example 6.6 has a minimum value of -1 and no maximum; thus its range consists of all $y \geq -1$.

EXAMPLE 6.7

Find the vertex, maximum or minimum, and range of

$$f(x) = 1.2x^2 + 3.7x - 5.9.$$

The x-coordinate of the vertex is

$$x = \frac{-b}{2a} = \frac{-3.7}{2(1.2)} = -1.54,$$

and its y-coordinate is

$$y = f(-1.54) = -8.75.$$

To decide whether the function has a maximum or a minimum, we need to know if the graph opens up or down. Here $a = 1.2 > 0$, so it opens up and the function has a minimum: this is the y-coordinate of the vertex, or -8.75.

The function has no maximum, so its range consists of all $y \geq -8.75$. See Figure 6.10. ▲

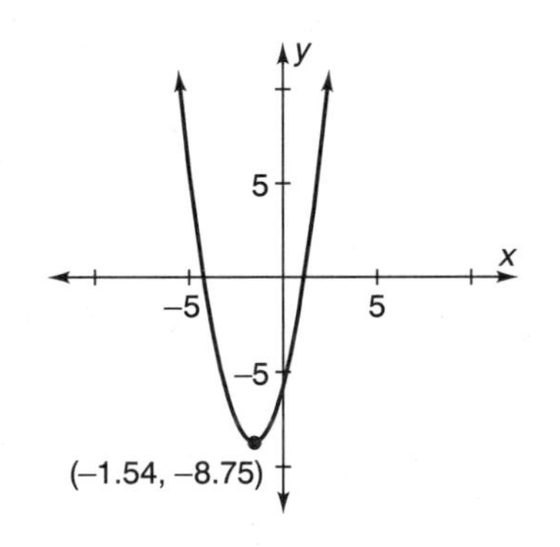

Figure 6.10

EXAMPLE 6.8

Find the vertex, maximum or minimum, and range of

$$f(x) = 61.5 - 3.4x - 5x^2.$$

The vertex of this parabola is $(-.34, 62.08)$; it opens down and has a maximum value of 62.08; and its range consists of all $y \leq 62.08$. See Figure 6.11. ▲

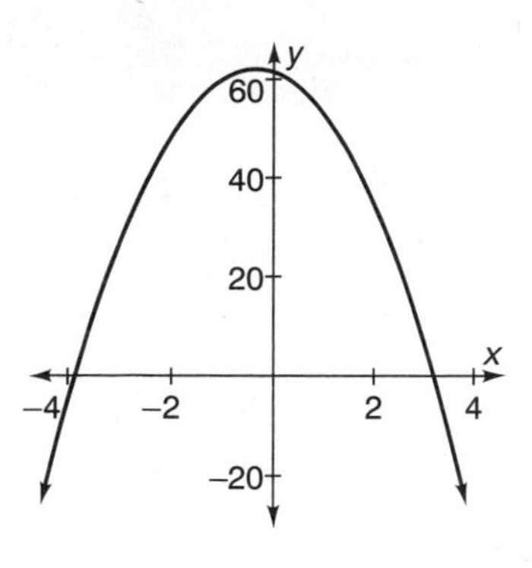

Figure 6.11

If a number k is in the range of a quadratic function, we can solve equations of the form $f(x) = k$. In Example 6.7 above, we can solve the equation $f(x) = 1.4$ since $1.4 \geq -8.75$. Replace $f(x)$ by 1.4 and solve

$$1.4 = 1.2x^2 + 3.7x - 5.9.$$

First put the equation in standard form

$$0 = 1.2x^2 + 3.7x - 7.3,$$

and use the quadratic formula to solve. Now $a = 1.2$, $b = 3.7$, and $c = -7.3$, giving us solutions $x = 1.37$ and $x = -4.45$. This tells you that you have two solutions to the equation

$$1.4 = 1.2x^2 + 3.7x - 5.9,$$

$x = 1.37$ and $x = -4.45$. See Figure 6.12.

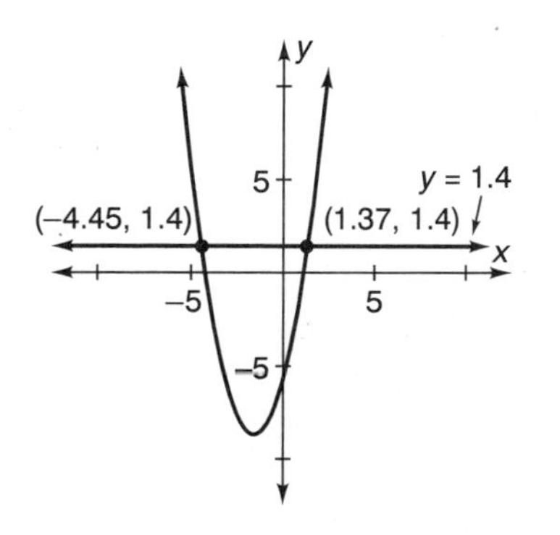

Figure 6.12

6.4 THINGS TO DO

Determine the maximum or minimum value and the range of each of these quadratic functions.

1. $f(x) = x^2 - 4x + 3$

2. $f(x) = .07 - 1.1x - .05x^2$

3. $H(t) = 110 + 4t^2$

4. $S(q) = 3q^2 - 6q$

5. $T(x) = 14x^2$

6. $R(x) = x^2 - 8x + 16$

7. $G(t) = t - 15 - 2.1t^2$

8. $h(x) = -x^2 + 1$

6.5 GRAPHS OF QUADRATIC FUNCTIONS

In the next three examples, 6.9, 6.10, and 6.11 we graph quadratic functions from start to finish, and provide all significant information. This information consists of vertex, intercepts, maximum or minimum, and range.

EXAMPLE 6.9

Graph $f(x) = -1.1x^2 + 30.2x - 197$.

First, $a = -1.1 < 0$, so the graph opens down. Its y-intercept is -197, which occurs when $x = 0$. To find its x-intercepts, solve the equation

$$0 = -1.1x^2 + 30.2x - 197$$

by using the quadratic formula.

$$x = \frac{-30.2 \pm \sqrt{(30.2)^2 - 4(-1.1)(-197)}}{2(-1.1)}$$

$$x = 10.67 \qquad \text{or} \qquad x = 16.78.$$

Next, we find its vertex. The x-coordinate is

$$x = \frac{-30.2}{2(-1.1)} = 13.73,$$

and the y-coordinate is

$$f(13.73) = 10.28.$$

This function has a maximum of 10.28, and its range consists of all $y \leq 10.28$. Figure 6.13 shows the graph. ▲

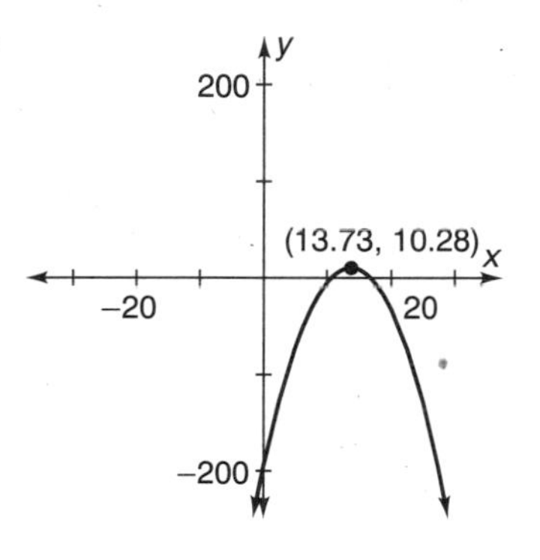

Figure 6.13

EXAMPLE 6.10

Graph $f(x) = 0.4x^2 - x + 7.3$ and determine the intercepts.

This one opens up because $a = 0.4 > 0$. Its y-intercept is 7.3, and x-intercepts are obtained as solutions to

$$0 = 0.4x^2 - x + 7.3.$$

Substitution into the quadratic formula gives

$$x = \frac{-(1.1) \pm \sqrt{(-1)^2 - 4(0.4)(7.3)}}{2(0.4)}.$$

The quantity under the radical is -10.68, a negative number. Hence this equation has no real solution; that is, the graph has no x-intercepts. (This does not mean there is no graph, only that it does not cross the x-axis.)

The vertex has coordinates

$$x = \frac{-(-1)}{2(0.4)} = 1.25,$$

and $y = f(1.25) = 6.68$. (Helpful hint: if we had first found the vertex, we would have known that there are no x-intercepts. The y-coordinate of the vertex is positive, and the graph opens up, so it is impossible for y to be zero anywhere on this graph!) This function has a minimum of 6.68, and its range consists of all numbers $y \geq 6.68$. See Figure 6.14. ▲

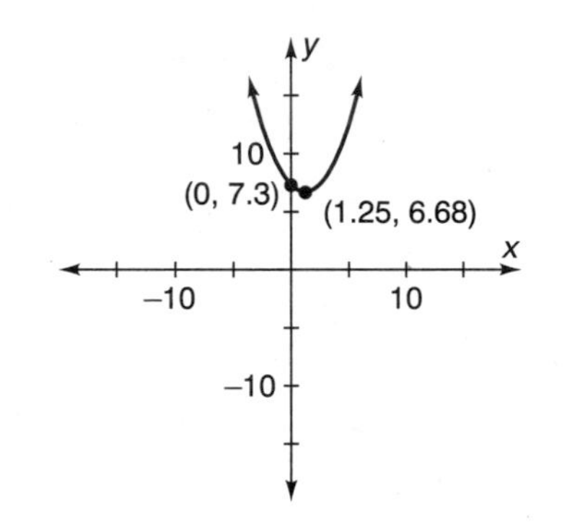

Figure 6.14

EXAMPLE 6.11

Graph $y = x^2 - 2.4x + 1.44$.

Here $a = 1 > 0$, so this parabola opens up. The y-intercept is the point (0, 1.44) and the x-intercepts are the solutions to the equation.

$$0 = x^2 - 2.4x + 1.44$$

$$x = \frac{-(-2.4) \pm \sqrt{(-2.4)^2 - 4(1)(1.44)}}{2(1)}$$

$$= \frac{2.4 \pm \sqrt{0}}{2} = 1.2$$

Hence there is only one x-intercept and it is the point (1.2, 0). This makes our work easy from here on. When there is only one x-intercept, the only possible place it can be is the vertex of the parabola. (See Figure 6.3.) Therefore the vertex of this parabola is (1.2, 0); its minimum is 0, and its range consists of all $y \geq 0$. (See Figure 6.15.) ▲

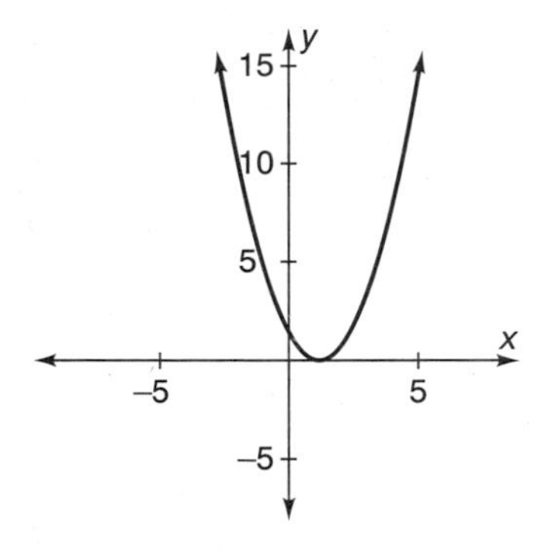

Figure 6.15

In the next two examples, we look at solutions to certain quadratic equations and what they mean relative to a graph.

EXAMPLE 6.12

Solve the equation

$$15 = x^2 - 2.4x + 1.44$$

for x.

First, put the equation in standard form by subtracting 15 from both sides.

$$0 = x^2 - 2.4x - 13.56$$

Then substitute coefficients into the quadratic formula to get solutions

$$x = 5.07 \qquad \text{or} \qquad x = -2.67 \text{ (rounded).}$$

Note that the graph of $y = x^2 - 2.4x + 1.44$ is shown in Figure 6.15. The solutions obtained in this example are the x-intercepts of the horizontal line $y = 15$ and the parabola $y = x^2 - 2.4x - 13.56$. Figure 6.16 shows both the line and the parabola with the x-coordinates indicated. ▲

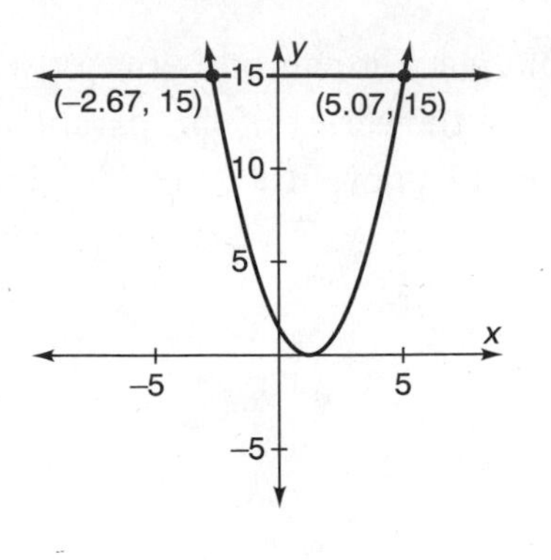

Figure 6.16

EXAMPLE 6.13

Solve the equation

$$5 = .4x^2 - x + 7.3.$$

In standard position, this equation is

$$0 = .4x^2 - x + 2.3.$$

Substitution of coefficients into the quadratic formula yields a negative under the radical, which means the equation has no real solution. The graph of $y = .4x^2 - x + 7.3$ is shown in Figure 6.14; the minimum of this function is 6.68 and hence the parabola can not possibly intersect the line $y = 5$. This situation is shown in Figure 6.17. ▲

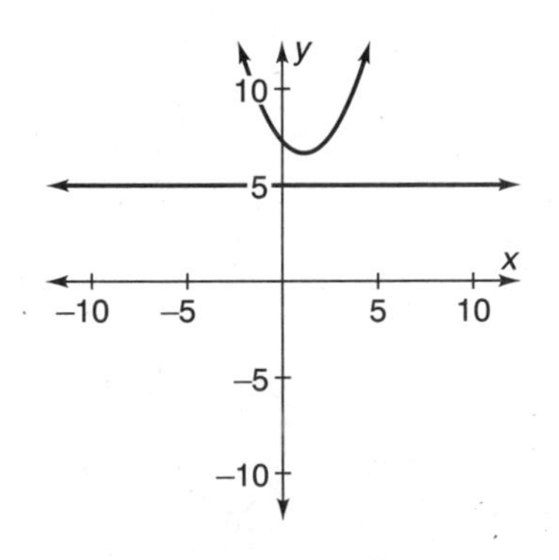

Figure 6.17

6.5 THINGS TO DO

Graph each quadratic function in Exercises 1–10, indicating intercepts, vertex, maximum or minimum, and range.

1. $y = x^2 - 6x - 10$
2. $y = 14 + x - 2x^2$
3. $y = 4 - 6x^2$
4. $y = 5.3x^2 + 2.3$
5. $y = 2.5x^2 - x + 7.3$
6. $f(x) = 8x^2$
7. $A(t) = .01t^2 - .14t$
8. $D(t) = t^2 + 4.6t + 5.29$
9. $S(t) = 2t - t^2 - 1$
10. $Q(t) = t^2 - 9.9$

Solve each of the following equations in Exercises 11–15, indicating the solution(s), if any, on a graph as illustrated in Examples 6.12 and 6.13.

11. $5 = x^2 - 6x - 10$. Refer to Exercise 1.
12. $12 = 14 + x - 2x^2$. Refer to Exercise 2.
13. $1 = 5.3x^2 + 2.3$. Refer to Exercise 4.
14. $2.5x^2 - x + 7.3 = -1$. Refer to Exercise 5.
15. $8x^2 = 100$. Refer to Exercise 6.

6.6 YOU AND YOUR CALCULATOR GRAPHS

Your calculator will graph just about any function. Also, it will estimate the coordinates of any point on the graph. But, before you begin your graph, you need to know certain information about the function. We provide you with illustrative examples.

EXAMPLE 6.14

Graph $f(x) = 0.9x^2 + 8.2x - 239.7$ on your calculator. Be sure that vertex and intercepts are visible. Also, find vertex coordinates and estimate x-intercepts with the calculator.

Enter the function and press **graph.** You probably don't see anything except axes. This is because you are looking at the wrong part of the xy-plane. To find a suitable area to see the important graph points, you need to do some work. First, calculate the vertex:

$$x = \frac{-8.2}{2(0.9)} = -4.56 \ (\textit{rounded}),$$

and

$$y = f(-4.56) = -258.38 \text{ (evaluate with calculator).}$$

This parabola opens up, so the entire graph will lie above $y = -258.38$, and there must be two x-intercepts. Set the y-range to y min $= -300$ (or smaller than -258.38) and y max $= 50$ (or any positive number that allows viewing the graph above the x-axis); then return to your graph. Next press the **trace** key and move the cursor along the graph to the right until it crosses the x-axis; set x max to be any number larger than the calculator estimate of this intercept (x max $= 20$). Now move the cursor to the left to estimate the smaller x-intercept, and set x min to be any number smaller than the calculator estimate, say x min $= -25$. Return to your graph, and you should see Figure 6.18. Finally, you can estimate the

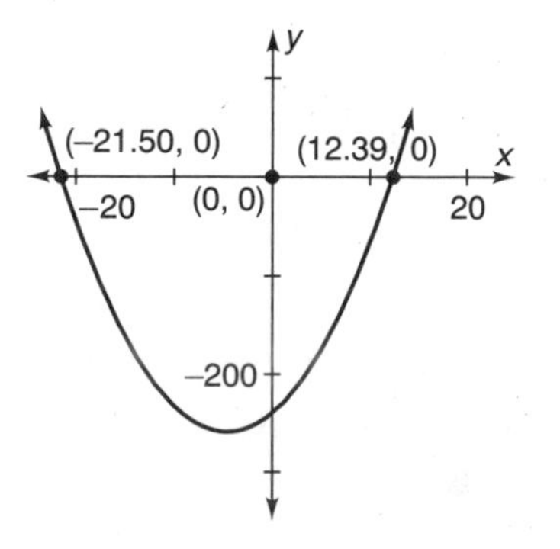

Figure 6.18

coordinates of the x-intercepts by positioning the cursor as close as possible to one intercept, and using the "zoom" feature of your calculator. The number $x = 12.39$ is a good estimate for the larger one, and $x = -21.50$ is a good estimate of the smaller. ▲

Actually, you can estimate the coordinates of any point on a graph using the "zoom" feature. For example, see how close you can estimate the vertex of this parabola using this technique.

EXAMPLE 6.15

Graph $f(x) = -0.03x^2 + .11x - 1.2$ on your calculator so that vertex and intercepts are visible. Also estimate coordinates of these important points.

Enter this function and press **graph.** Unless your x range is very small, you should see the vertex on your screen. The graph opens down, and the vertex is below the x-axis, so there are no x-intercepts. You can use your calculator to estimate the coordinates of the vertex. With the **trace** key, position the cursor on the graph as close as possible to the vertex, and use the "zoom" feature. You will know when the cursor is near the vertex if the y-coordinates shown on the screen increase, then decrease as the cursor moves from left to right. See Figure 6.19. A good estimate is $(1.87, -1.099)$. ▲

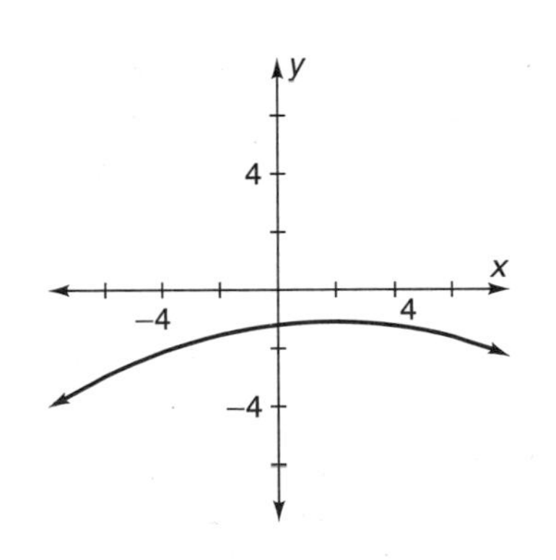

Figure 6.19

6.6 THINGS TO DO

Graph the quadratic functions in Exercises 1–4 using the calculator. Use the **trace** *key on the calculator to arrive at good estimates for the vertex and any other points you are asked to find.*

1. $HD(x) = .0006x^2 - .126x + 12.9.$
2. $Joe(y) = -3.2(y^2 - 1.1y + 4) + 5.$
3. $K(b) = .003b^2 - 1.09b + 21.2$; when is $K(b) = 45$? 50?
4. $L(t) = -12.2t^2 + 54.6t - 1.6$; when is $L(t) = 30$? 40? 50? 60?

Chapter 7

Systems of Linear Equations and Matrices

7.1 SYSTEMS OF LINEAR EQUATIONS

A linear equation in n variables can be written in the form

$$a_1x_1 + a_2x_2 + \cdots + a_nx_n = c,$$

where $x_1, x_2, \ldots x_n$ are the variables and $a_1, a_2, \ldots a_n$, c are constants. A system of m linear equations in n variables can be written in this form:

$$\begin{aligned}
a_{11}x_1 + a_{12}x_2 + \cdots + a_{1n}x_n &= c_1 \\
a_{21}x_1 + a_{22}x_2 + \cdots + a_{2n}x_n &= c_2 \\
&\vdots \\
a_{m1}x_1 + a_{m2}x_2 + \cdots + a_{mn}x_n &= c_m,
\end{aligned}$$

Note: Chapter 7 is a prerequisite for Chapters 8 and 9.

where, once again, $x_1, \ldots, x_n$ are the variables, and a_{ij}, $i = 1, 2, \ldots, m$, $j = 1, 2, \ldots, n$, and c_k, $k = 1, \ldots, m$, are the constants. This is called an $m \times n$ system. Notice that the constant a_{ij} is the coefficient of the jth variable x_j in the ith equation; for example, a_{34} is the coefficient of x_4 in the third equation. These subscripts will be important later.

A solution to a system of linear equations is a set of n numbers

$$x_1 = b_1, x_2 = b_2, \ldots, x_n = b_n$$

which make all the equations true.

We focus on methods of finding solutions to 2×2 and 3×3 systems. The system

$$\begin{aligned} x + y &= 5 \\ 3x + 4y &= 1 \end{aligned}$$

is a 2×2 system, whereas

$$\begin{aligned} x + 2y + z &= 0 \\ 2x - y - 3z &= 1 \\ 3x - 2y &= -1 \end{aligned}$$

is a 3×3 system.

When solving a system of linear equations, the following three operations are used.

1. Interchange two equations.
2. Replace any equation by a nonzero multiple of that equation.
3. Replace any equaton by itself plus a multiple of another.

Solving a system of equations using these operations is called solving by elimination of variables.

EXAMPLE 7.1

Solve the 2×2 system

$$\begin{aligned} x + y &= 5 \\ 3x - 4y &= 1. \end{aligned}$$

The variable y can be eliminated by replacing the second equation by itself plus 4 times the first equation:

$$\begin{aligned} 4(x + \quad y &= 5) \\ +(3x - 4y &= 1), \end{aligned}$$

or

$$\begin{aligned} 4x + 4y &= 20 \\ +(3x - 4y &= 1), \end{aligned}$$

so the new second equation is

$$7x = 21,$$

and hence

$$x = 3.$$

To find y, substitute $x = 3$ into either of the original equations; if we choose the first, we get

$$3 + y = 5,$$

so

$$y = 2.$$

The solution to this system is $x = 3$, $y = 2$.

Geometrically, this solution gives the coordinates of the point of intersection of the graph of the two equations. See Figure 7.1. ▲

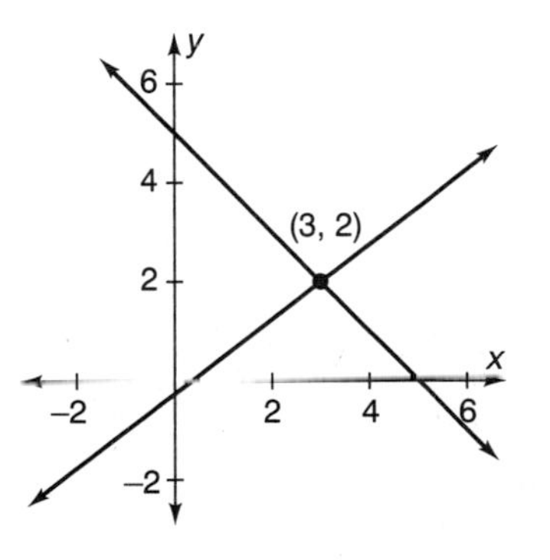

Figure 7.1

EXAMPLE 7.2

Solve the system

$$\begin{aligned} 3x - y &= 4 \\ -6x + 2y &= 10. \end{aligned}$$

We eliminate x in the second equation by multiplying the first equation by 2 and adding to the second to get

$$0 = 18.$$

Everyone knows that 0 can't equal 18. This means that this system has no solution. Geometrically the lines which are the graphs of these two equations are parallel. See Figure 7.2. ▲

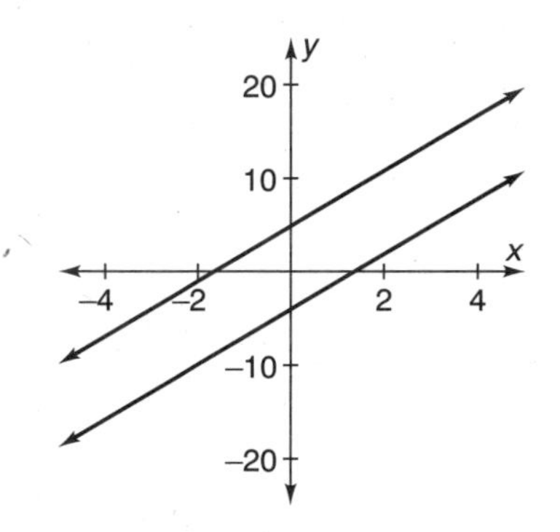

Figure 7.2

EXAMPLE 7.3

Solve the 3×3 system

$$\begin{aligned} x - y + z &= 8 \\ x + 2y - z &= -5 \\ 2x + 3y + z &= 6. \end{aligned}$$

There's more to do here because there are more variables. First we eliminate z from the second equation by adding the first and second to get

$$2x + y = 3.$$

To get another equation in x and y, eliminate z from the third equation by adding the third and (-1) times the first; this gives

$$x + 4y = -2.$$

Now, we can solve the resulting 2×2 system,

$$\begin{aligned} 2x + y &= 3 \\ x + 4y &= -2 \end{aligned}$$

by again eliminating a variable. Eliminate x by adding the first to -2 times the second to get

$$-7y = 7,$$

so

$$y = -1.$$

Substitute this $y = -1$ into $x + 4y = -2$ (or into $2x + y = 3$) to get

$$x - 4 = -2,$$

or

$$x = 2.$$

Finally substitute $x = 2$ and $y = -1$ into any one of the original three equations to find z. If we choose the first, we get

$$2 - (-1) + z = 8,$$

so

$$z = 5.$$

Therefore, the solution is

$$x = 2, y = -1, z = 5.$$

(You can check these answers by substituting into the original equations.) ▲

EXAMPLE 7.4

Solve the system

$$\begin{aligned} x + 2y + z &= 4 \\ x + 3y - 5z &= -2 \\ 2x - y - 2z &= 4. \end{aligned}$$

We eliminate x from the second and third equation to get a 2×2 system in y and z. First, subtract the first from the second equation to get

$$y - 6z = -6.$$

Next subtract twice the first from the third to get

$$-5y - 4z = -4.$$

Now solve the 2×2 system

$$\begin{aligned} y - 6z &= -6 \\ -5y - 4z &= -4 \end{aligned}$$

to get

$$y = 0, z = 1.$$

Substitute these variables into the first original equation (or any other one) to see that

$$x = 3. \quad \blacktriangle$$

7.1 THINGS TO DO

Solve the following systems of equations by elimination of variables.

1. $\begin{aligned} x + 3y &= 1 \\ 2x - 3y &= 2 \end{aligned}$

2. $\begin{aligned} 4x + y &= 5 \\ 8x + 3y &= 2 \end{aligned}$

3. $\begin{aligned} x - y &= 1 \\ 2x + 3y &= -2 \end{aligned}$

4. $\begin{aligned} x - y &= 3 \\ 2x + 3y &= 1 \end{aligned}$

5. $\begin{aligned} 4x + y &= 1 \\ 2x + 3y &= -2 \end{aligned}$

6. $\begin{aligned} x + y &= 6 \\ y + z &= 10 \\ x + z &= 18 \end{aligned}$

7. $\begin{aligned} x + 2y + z &= 3 \\ x + 3y + 4z &= 6 \\ 2x + 4y + 3z &= 7 \end{aligned}$

8. $\begin{aligned} 2x + y + z &= 7 \\ 2x + y - z &= 5 \\ 2x + 4y + 3z &= 12 \end{aligned}$

9. $\begin{aligned} 2x + y + 3z &= 1 \\ 3x - y - 2z &= 3 \\ 4x - y + 3z &= 5 \end{aligned}$

10. $\begin{aligned} 2x + 3y - z &= 0 \\ 3x - 2y + z &= 7 \\ 4x + 6y - 3z &= 5 \end{aligned}$

7.2 MATRICES

This is what you need to know about matrices for *Earth Algebra:* a matrix is a rectangular array of numbers. Its size is designated by the number of rows and columns it has; for example a matrix with four rows and six columns is a 4×6 matrix. The numbers in the matrix are called entries; the number in the ith row and j^{th} column is the i, j-entry; for example, the number in the second row and third column is the 2,3-entry. The general form for an $m \times n$ matrix is

$$\begin{bmatrix} a_{11} & a_{12} & \cdots & a_{1n} \\ a_{21} & a_{22} & \cdots & a_{2n} \\ \cdot & \cdot & \cdot & \cdot \\ a_{m1} & a_{m2} & a_{m3} & a_{mn} \end{bmatrix},$$

where a_{ij} is the i, j-entry.

An example of a 3×4 matrix is

$$\begin{bmatrix} -1 & -6 & 8 & -2 \\ 0 & 7 & 14 & 10 \\ 2 & -3 & 0 & -3 \end{bmatrix}$$

Its 2,3-entry is $a_{23} = 14$; its 3,1-entry is $a_{31} = 2$, and so on.

Matrices of the same size can be added together or subtracted from each other by adding or subtracting corresponding entries. (You can not add or subtract matrices of different sizes.) We give an example of each of these operations.

EXAMPLE 7.5

$$\begin{bmatrix} 1 & 0 & 4 \\ -2 & 6 & -1 \end{bmatrix} + \begin{bmatrix} 5 & 1 & -9 \\ 10 & 2 & 3 \end{bmatrix} = \begin{bmatrix} 6 & 1 & -5 \\ 8 & 8 & 2 \end{bmatrix}. \quad \blacktriangle$$

EXAMPLE 7.6

$$\begin{bmatrix} 3 & 1 & 2 \\ 0 & -1 & 4 \\ 7 & 0 & 8 \end{bmatrix} - \begin{bmatrix} 5 & 2 & -3 \\ 4 & 0 & 6 \\ 1 & -1 & 2 \end{bmatrix} = \begin{bmatrix} -2 & -1 & 5 \\ -4 & -1 & -2 \\ 6 & 1 & 6 \end{bmatrix}. \quad \blacktriangle$$

A matrix can also be multiplied by any real number; multiply each entry by the real number. This operation is called scalar multiplication.

EXAMPLE 7.7

$$3\begin{bmatrix} -1 & 0 \\ 7 & -4 \\ 2 & 8 \end{bmatrix} = \begin{bmatrix} -3 & 0 \\ 21 & -12 \\ 6 & 24 \end{bmatrix}. \quad \blacktriangle$$

Another very important operation on matrices is matrix multiplication. If two matrices are the right size, then they can be multiplied together to produce a new matrix. If A and B are matrices, in order to perform the product AB, the matrix A must have the same number of columns as the number of rows of the matrix B. For the purpose of illustration, this first example is simple.

EXAMPLE 7.8

$$\begin{bmatrix} 1 & 2 & 3 \end{bmatrix}\begin{bmatrix} 4 \\ 5 \\ 6 \end{bmatrix} = [1 \cdot 4 + 2 \cdot 5 + 3 \cdot 6] = [32].$$

Here $A = \begin{bmatrix} 1 & 2 & 3 \end{bmatrix}$ is 1×3, and $B = \begin{bmatrix} 4 \\ 5 \\ 6 \end{bmatrix}$ is 3×1, and their product $AB = [32]$ is 1×1. ▲

Example 7.8 shows how to multiply a $1 \times k$ row matrix by a $k \times 1$ column matrix: just multiply corresponding entries and sum. In general, if A is $m \times k$ and B is $k \times n$, then the product AB is $m \times n$. This product matrix is defined by multiplying the ith row of A by the jth column of B to produce the i,j-entry of AB (as in Example 7.8).

EXAMPLE 7.9

$$\begin{aligned} &\begin{bmatrix} 1 & -2 \\ 4 & -3 \end{bmatrix}\begin{bmatrix} 4 & -1 & 0 \\ 0 & 2 & 5 \end{bmatrix} \\ &\quad = \begin{bmatrix} 1 \cdot 4 + (-2)0 & 1(-1) + (-2)2 & 1 \cdot 0 + (-2)5 \\ 4 \cdot 4 + (-3)0 & 4(-1) + (-3)2 & 4 \cdot 0 + (-3)5 \end{bmatrix} \\ &\quad = \begin{bmatrix} 4 & -5 & -10 \\ 16 & -10 & -15 \end{bmatrix}. \quad \blacktriangle \end{aligned}$$

EXAMPLE 7.10

$$\begin{bmatrix} 1 & 2 \\ 3 & 4 \end{bmatrix} \begin{bmatrix} 1 & 2 & 3 \\ 2 & 1 & 5 \end{bmatrix} = \begin{bmatrix} 5 & 4 & 13 \\ 11 & 10 & 29 \end{bmatrix}. \quad \blacktriangle$$

EXAMPLE 7.11

$$\begin{bmatrix} 1 & 2 & 3 \\ 4 & 5 & 6 \\ 7 & 8 & 9 \end{bmatrix} \cdot \begin{bmatrix} 1 & 0 & 0 \\ 0 & 1 & 0 \\ 0 & 0 & 1 \end{bmatrix} = \begin{bmatrix} 1 & 2 & 3 \\ 4 & 5 & 6 \\ 7 & 8 & 9 \end{bmatrix},$$

and

$$\begin{bmatrix} 1 & 0 & 0 \\ 0 & 1 & 0 \\ 0 & 0 & 1 \end{bmatrix} \cdot \begin{bmatrix} 1 & 2 & 3 \\ 4 & 5 & 6 \\ 7 & 8 & 9 \end{bmatrix} = \begin{bmatrix} 1 & 2 & 3 \\ 4 & 5 & 6 \\ 7 & 8 & 9 \end{bmatrix}. \quad \blacktriangle$$

EXAMPLE 7.12

$$\begin{bmatrix} 2 & 5 \\ 1 & 3 \end{bmatrix} \cdot \begin{bmatrix} 3 & -5 \\ -1 & 2 \end{bmatrix} = \begin{bmatrix} 1 & 0 \\ 0 & 1 \end{bmatrix},$$

and

$$\begin{bmatrix} 3 & -5 \\ -1 & 2 \end{bmatrix} \cdot \begin{bmatrix} 2 & 5 \\ 1 & 3 \end{bmatrix} = \begin{bmatrix} 1 & 0 \\ 0 & 1 \end{bmatrix}. \quad \blacktriangle$$

In Example 7.11 the matrix with all the 0s and 1s is special. It is called the identity matrix. The $n \times n$ identity matrix has 1's on its diagonal, and 0's elsewhere. If A and B are matrices and I is an identity matrix such that the products AI and IB are defined, then $AI = A$ and $IB = B$.

In Example 7.12, the matrix

$$A = \begin{bmatrix} 2 & 5 \\ 1 & 3 \end{bmatrix}$$

has what is known as an inverse matrix. It is

$$A^{-1} = \begin{bmatrix} 3 & -5 \\ -1 & 2 \end{bmatrix}.$$

From this same example, you see that

$$AA^{-1} = I(2 \times 2 \text{ identity}),$$

and also that

$$A^{-1}A = I.$$

In general, an $n \times n$ matrix has an inverse if there is an $n \times n$ matrix $B = A^{-1}$ such that

$$AB = BA = I(n \times n \text{ identity}).$$

Only square matrices ($n \times n$) have inverses, and not all of these do. You will see in the next section the usefulness of matrices that have inverses. Determining the inverse of a matrix (if it has one) by hand is beyond the scope of this book, but you'll see how to use your calculator to find the inverses of the matrices in *Earth Algebra*.

7.2 THINGS TO DO

Perform the indicated operation whenever it is defined.

1. $\begin{bmatrix} 3 & 1 \\ 2 & 6 \\ 1 & 3 \end{bmatrix} + \begin{bmatrix} 0 & 2 \\ 1 & -1 \\ 7 & 5 \end{bmatrix}$

2. $\begin{bmatrix} 3 & 1 \\ 2 & 6 \\ 1 & 3 \end{bmatrix} - \begin{bmatrix} 0 & 2 \\ 1 & -1 \\ 7 & 5 \end{bmatrix}$

3. $\begin{bmatrix} 1 & 3 \\ 2 & 4 \end{bmatrix} - \begin{bmatrix} 4 & 3 \\ 2 & 1 \end{bmatrix}$

4. $\begin{bmatrix} 1 & 0 \\ 0 & 1 \end{bmatrix} + \begin{bmatrix} 0 & 0 & 0 \\ 0 & 0 & 0 \end{bmatrix}$

5. $5\begin{bmatrix} 2 & 1 \\ 3 & 6 \end{bmatrix}$

6. $-2\begin{bmatrix} 1 & 1 & 1 \\ 3 & 2 & 5 \end{bmatrix}$

7. $2\begin{bmatrix} 3 & 1 \\ 2 & 0 \end{bmatrix} - 4\begin{bmatrix} 0 & 2 \\ 1 & 3 \end{bmatrix}$

8. $[0 \quad 0] + [3 \quad 7]$

9. $[1 \quad 3 \quad 2] + \begin{bmatrix} 2 \\ 3 \\ 7 \end{bmatrix}$

10. $[3 \quad 1]\begin{bmatrix} 2 \\ 5 \end{bmatrix}$

11. $\begin{bmatrix} 1 & 2 \\ 3 & 1 \end{bmatrix}\begin{bmatrix} 5 \\ 2 \end{bmatrix}$

12. $\begin{bmatrix} 2 \\ 5 \end{bmatrix}[3 \quad 1]$

13. $[3 \quad 2 \quad 4]\begin{bmatrix} 1 \\ 3 \end{bmatrix}$

14. $\begin{bmatrix} 1 & 6 & 2 \\ 6 & 1 & 4 \\ -1 & 2 & 1 \end{bmatrix}\begin{bmatrix} 3 \\ 2 \\ -6 \end{bmatrix}$

15. $\begin{bmatrix} 2 & 3 \\ 1 & 5 \end{bmatrix}\begin{bmatrix} 2 \\ 1 \end{bmatrix}$

16. $\begin{bmatrix} 23 & 29 \\ 17 & 43 \end{bmatrix}\begin{bmatrix} 1 & 0 \\ 0 & 1 \end{bmatrix}$

17. $\begin{bmatrix} 1 & 0 & 0 \\ 0 & 1 & 0 \\ 0 & 0 & 1 \end{bmatrix}\begin{bmatrix} 2 & 3 & 7 \\ 3 & 1 & 2 \\ 1 & 5 & 8 \end{bmatrix}$

18. $\begin{bmatrix} 2 & 3 \\ 1 & 5 \\ 4 & 2 \end{bmatrix}\begin{bmatrix} 1 & 0 \\ 0 & 1 \end{bmatrix}$

19. $\begin{bmatrix} 1 & 0 \\ 0 & 1 \end{bmatrix}\begin{bmatrix} 2 & 3 \\ 1 & 5 \\ 4 & 2 \end{bmatrix}$

In Exercises 20–22, determine whether the following pairs of matrices are inverses of each other.

20. $\begin{bmatrix} 3 & 1 \\ 2 & 1 \end{bmatrix}, \begin{bmatrix} 1 & -1 \\ -2 & 3 \end{bmatrix}$

21. $\begin{bmatrix} 1 & 3 \\ 2 & 6 \end{bmatrix}, \begin{bmatrix} 4 & -3 \\ -1 & 1 \end{bmatrix}$

22. $\begin{bmatrix} 1 & 2 & 3 \\ 2 & 5 & 6 \end{bmatrix}, \begin{bmatrix} -1 & 1 \\ -2 & 1 \\ 2 & -1 \end{bmatrix}$

7.3 SYSTEMS OF EQUATIONS AND MATRICES

In this section, we illustrate the relationship between matrices and systems of equations, and how a system of linear equations can sometimes be solved using matrix methods.

Consider the system of 3 equations in 3 variables,

$$\begin{aligned} 4x + y - 2z &= 7 \\ -3x - y + 2z &= 3 \\ 5x + y - 3z &= 4. \end{aligned}$$

We make a matrix of coefficients for the unknowns x, y, and z. There will be one row for each equation, and one column for each unknown. The first row contains coefficients of the first equation; for example, the 1,1-entry is 4, which is the coefficient of the first unknown x in the first equation. The 1,2-entry is 1, which is the coefficient of the second unknown y in the first equation. To complete the first row, the 1,3-entry is -2. Here's the matrix, called the coefficient matrix of the system:

$$A = \begin{bmatrix} 4 & 1 & -2 \\ -3 & -1 & 2 \\ 5 & 1 & -3 \end{bmatrix}.$$

This is a 3×3 matrix. Its 3,1-entry is 5, and 5 is the coefficient of x in the third equation.

The other numbers in this system of equations are the constants on the right hand side of the equal sign; put them in a 3×1 matrix, which is called the matrix of constants for the system:

$$B = \begin{bmatrix} 7 \\ 3 \\ 4 \end{bmatrix}.$$

Notice that there is one row for each equation and the entry in each row is the constant in the corresponding equation. Next, put the unknowns in another 3×1 matrix

$$X = \begin{bmatrix} x \\ y \\ z \end{bmatrix}.$$

The product

$$AX = B$$

produces the original system of equations:

$$\begin{bmatrix} 4 & 1 & -2 \\ -3 & -1 & 2 \\ 5 & 1 & -3 \end{bmatrix} \begin{bmatrix} x \\ y \\ z \end{bmatrix} = \begin{bmatrix} 4x + y - 2z \\ -3x - y + 2z \\ 5x + y - 3z \end{bmatrix} = \begin{bmatrix} 7 \\ 3 \\ 4 \end{bmatrix}.$$

The coefficient matrix A in this example does have an inverse

$$A^{-1} = \begin{bmatrix} 1 & 1 & 0 \\ 1 & -2 & -2 \\ 2 & 1 & -1 \end{bmatrix}.$$

If we multiply both sides of the matrix equation $AX = B$ by A^{-1}, here's what happens:

$$\begin{bmatrix} 1 & 1 & 0 \\ 1 & -2 & -2 \\ 2 & 1 & 1 \end{bmatrix} \begin{bmatrix} 4 & 1 & -2 \\ -3 & -1 & 2 \\ 5 & 1 & -3 \end{bmatrix} \begin{bmatrix} x \\ y \\ z \end{bmatrix} = \begin{bmatrix} 1 & 1 & 0 \\ 1 & -2 & -2 \\ 2 & 1 & 1 \end{bmatrix} \begin{bmatrix} 7 \\ 3 \\ 4 \end{bmatrix}$$

$$\begin{bmatrix} 1 & 0 & 0 \\ 0 & 1 & 0 \\ 0 & 0 & 1 \end{bmatrix} \begin{bmatrix} x \\ y \\ z \end{bmatrix} = \begin{bmatrix} 10 \\ -7 \\ 13 \end{bmatrix}$$

so

$$\begin{bmatrix} x \\ y \\ z \end{bmatrix} = \begin{bmatrix} 10 \\ -7 \\ 13 \end{bmatrix},$$

which means the solution to the original system is

$$x = 10, y = -7, z = 13.$$

What you have just seen always works as long as the coefficient matrix A is square ($n \times n$) and has an inverse. That is, if A is the coefficient matrix of a system of n equations in n variables with inverse

A^{-1}, and if X is the column matrix consisting of the n variables, and B is the column matrix consisting of the n constants, then the original system is represented by the matrix equation

$$AX = B;$$

multiply both sides of this equation by A^{-1} to determine the solution matrix

$$X = A^{-1}B.$$

EXAMPLE 7.13

Solve the system of equations

$$\begin{aligned} 3x + 3y - z &= 11 \\ -2x - 2y + z &= 4 \\ -4x - 5y + 2z &= 2 \end{aligned}$$

using matrix methods.

The coefficient matrix is

$$A = \begin{bmatrix} 3 & 3 & -1 \\ -2 & -2 & 1 \\ -4 & -5 & 2 \end{bmatrix};$$

the matrix of variables is

$$X = \begin{bmatrix} x \\ y \\ z \end{bmatrix};$$

and the matrix of constants is

$$B = \begin{bmatrix} 11 \\ 4 \\ 2 \end{bmatrix}.$$

We give you the inverse of A:

$$A^{-1} = \begin{bmatrix} 1 & -1 & 1 \\ 0 & 2 & -1 \\ 2 & 3 & 0 \end{bmatrix}.$$

Now, the solution is determined by the product

$$X = A^{-1}B = \begin{bmatrix} 1 & -1 & 1 \\ 0 & 2 & -1 \\ 2 & 3 & 0 \end{bmatrix} \begin{bmatrix} 11 \\ 4 \\ 2 \end{bmatrix} = \begin{bmatrix} 9 \\ 6 \\ 34 \end{bmatrix},$$

so

$$x = 9, y = 6, z = 34. \quad \blacktriangle$$

Your calculator will solve a system of equations for you. Here's how: first, set the dimensions of A and B. You do not need to enter X; the product $A^{-1}B$ will be X. Next, enter the matrices A and B, and finally, perform the product $A^{-1}B$ to see the solution matrix. (This method only works if A has an inverse.) There might be even easier ways to solve systems of equations on your calculator. Read the manual for more information.

EXAMPLE 7.14

Solve the system of equations

$$\begin{aligned} x + 5y - z &= 3 \\ 4x + 3y - 2z &= 5 \\ 2x - 2y + z &= 9 \end{aligned}$$

using your calculator.

The coefficient matrix is 3×3,

$$A = \begin{bmatrix} 1 & 5 & -1 \\ 4 & 3 & -2 \\ 2 & -2 & 1 \end{bmatrix};$$

the constant matrix is 3×1,

$$B = \begin{bmatrix} 3 \\ 5 \\ 9 \end{bmatrix}.$$

Set dimensions of each; enter each matrix, and take the product to get

$$A^{-1}B = \begin{bmatrix} 3 \\ 1 \\ 5 \end{bmatrix},$$

so $x = 3$, $y = 1$, and $z = 5$. ▲

EXAMPLE 7.15

Solve the system using your calculator,

$$\begin{aligned} 4x + y &= 7 \\ 2x + 3y &= -1. \end{aligned}$$

The coefficient matrix is 2×2,

$$A = \begin{bmatrix} 4 & 1 \\ 2 & 3 \end{bmatrix};$$

the constant matrix is 2×1,

$$B = \begin{bmatrix} 7 \\ -1 \end{bmatrix}.$$

The product matrix is

$$A^{-1}B = \begin{bmatrix} 2.2 \\ -1.8 \end{bmatrix},$$

so

$$x = 2.2, \; y = -1.8. \quad ▲$$

7.3 THINGS TO DO

Solve the following systems of equations using your calculator.

1. $x + y = 2$
   $3x + 2y = 5$

2. $3x + 7y = 4$
   $x + 4y = 5$

3. $7x + y = 1$
   $3x + 4y = -1$

4. $11x + y = 3$
   $3x + y = 7$

5. $3x + 2y = 8$
   $x + 4y = 5$

6. $15x + 2y = 10$
   $21x + 4y = -13$

7. $8x + 3y = 17$
   $-11x + 2y = 3$

8. $x - 3y - 2z = 11$
   $2x + 8y + 9z = 23$
   $x + 5y + 6z = 13$

9. $x - y + 3z = 2$
   $3x - y + 7z = 3$
   $x + y + 6z = -5$

10. $x + 7y + 4z = -3$
    $-3x - y + 16z = 7$
    $2x + 4y - z = 1$

11. $x + 3y + 7z = 4$
    $3x + 4y + 7z = 5$
    $2x + 6y + 9z = -8$

12.
$$\begin{aligned} 5x + 2y - z &= 4 \\ 4x - 3y + 2z &= 5 \\ 3x + 7y - 5z &= 8 \end{aligned}$$

13.
$$\begin{aligned} 100x + 10y + z &= 13.9 \\ 400x + 20y + z &= 13.5 \\ 900x + 30y + z &= 13.4 \end{aligned}$$

14.
$$\begin{aligned} x + 2z &= -1 \\ 2x + y &= 14 \\ 3y + 5z &= -3 \end{aligned}$$

CHAPTER 8

Modeling Carbon Dioxide Emission from Autos in the United States

8.1 PREDICTING TOTAL PASSENGER CARS IN THE FUTURE

Table 8.1 on the next page shows the number of U.S. passenger cars (in millions) in the indicated year.

TABLE 8.1 Number of Passenger Cars by Year

Years	Cars $\times 10^6$
1940	27.5
1950	40.3
1960	61.7
1970	89.3
1980	121.6
1986	135.4
1991	143.0

Source: *Statistical Abstract of the United States, 1993,* Bureau of the Census.

If we use t for the number of years after 1940, and $A(t)$ for the number of cars ($\times 10^6$) in year $1940 + t$, and plot this data on a (t, A)-coordinate system it looks like these points fall very nearly in a straight line, so we will model this data with a linear function. See Figure 8.1.

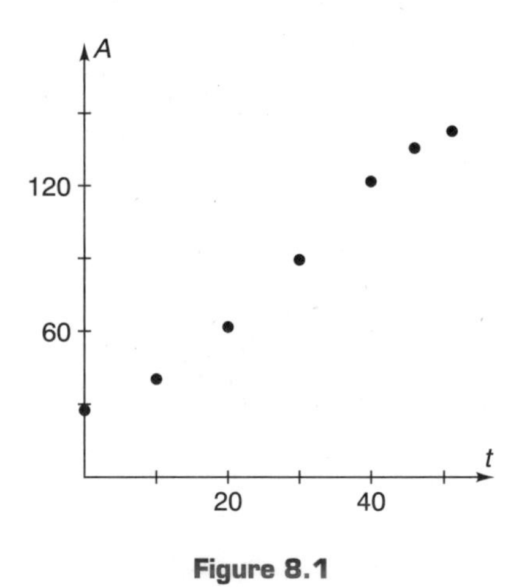

Figure 8.1

You need the best fitting line. Altogether there are twenty-one possible lines; your instructor will assign certain years to your group, and your group is to find the best line using points corresponding to these years. Then, each group should report its result, and the best line for this model can be determined.

Now, here are some helpful hints on making these reports.

The first, and most important, thing to remember is this: when you explain what you have done, almost everyone should be able to understand, even if they've not had college algebra. They may not be able to understand the derivation of the equation of the line (or whatever curve), but they should know why you are deriving it, what you intend to do with it, and how it works! More specifically, say what it is that you are modeling; graph the given data; say which years or points you are using to write the equation; show the graph of the equation with the points in the correct position relative to the line (or curve); say what the error is and how your group got it; and finally, show how your equation can be used to predict the future. (You could also report any additional interesting information your group's equation provides.)

Use your best equation to predict the following information:

1. the number of automobiles in the U.S. in the year 2015;
2. the year in which there will be 200 million automobiles in the U.S.;
3. how many more automobiles there are each year in the U.S.

EARTH NOTE Oil Consumptions

Did you know that transportation uses 60 percent of all oil consumed in the United States?

Source: Kaufman, Donald, and Franz, Cecelia, *Biosphere 2000: Protecting our Global Environment,* HarperCollins College Publishers (New York, 1994).

8.2 PREDICTING FUEL EFFICIENCY OF PASSENGER CARS

The next factor that we need to complete our model of CO_2 emission from automobiles is the amount of gasoline they use. The data below provide the average gasoline efficiency per car. Table 8.2 shows the average miles per gallon (MPG) of all U.S. automobiles in the indicated years.

TABLE 8.2
Average MPG By Year

Year	MPG
1940	14.8
1950	13.9
1960	13.4
1970	13.5
1980	15.5
1986	18.3
1991	21.7

Source: *Statistical Abstract of the United States, 1993,* Bureau of the Census.

This model is different, so we'll work through the four steps to modeling together. Steps 1 and 2 are easy; Figure 8.2 shows a (t, MPG) coordinate system with the data points plotted (again $t = 0$ in 1940). Look at the graph and think about what type of curve will best fit this data. After a moment's reflection, you should all agree that these points fall in a very parabolic pattern. Therefore, we will use a quadratic function to model this data, and then write the equation of the best parabola which fits these points. You may not know how to do this, so think about it for a minute. In particular, think back to your linear equations: you need two points to write the equation of a line, and any two points uniquely determine that line. (There's only one way to draw a line through two points.) There are many ways to draw parabolas through only two points,

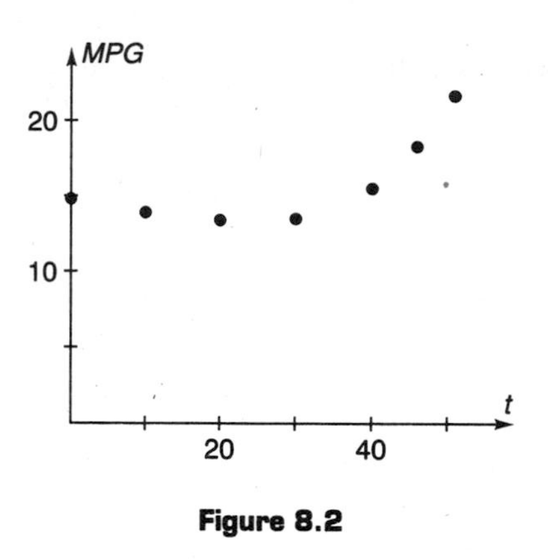

Figure 8.2

but there's at most one way to draw a parabola through three points. Given any three points which are not collinear, there is a unique parabola which passes through them. However, if three points are on the same line, no parabola can pass through all three of them. Any three points in Figure 8.2 will do. Eventually you'll need to consider all possible triples of data points to get all possible parabolas for this model. We first (somewhat arbitrarily) choose points corresponding to the years 1940, 1970, and 1986, or (0, 14.8), (30, 13.5), and (46, 18.3). To write the quadratic equation whose graph passes through these three points, you need the standard form

$$y = ax^2 + bx + c$$

or, using variables appropriate to our model, the function looks like

$$MPG(t) = at^2 + bt + c,$$

where $MPG(t)$ = average miles per gallon in year 1940 + t. You must always state what your variables mean so that anyone will understand. We'll work out two equations for you, and we put subscripts on *MPG* to distinguish them.

If the graph is to go through the chosen points, then:

$$\begin{aligned} \text{when } t &= 0, MPG_1(0) = 14.8; \\ \text{when } t &= 30, MPG_1(30) = 13.5; \\ \text{and when } t &= 46, MPG_1(46) = 18.3. \end{aligned}$$

When substituted into the standard form, you get

$$\begin{aligned}
14.8 &= a(0)^2 + b(0) + c, \\
13.5 &= a(30)^2 + b(30) + c, \\
18.3 &= a(46)^2 + b(46) + c,
\end{aligned}$$

so you have to solve the system of three linear equations in the unknowns a, b, and c:

$$\begin{aligned}
14.8 &= c \\
13.5 &= 900a + 30b + c \\
18.3 &= 2116a + 46b + c.
\end{aligned}$$

The coefficient matrix is

$$A = \begin{bmatrix} 0 & 0 & 1 \\ 900 & 30 & 1 \\ 2116 & 46 & 1 \end{bmatrix}$$

and the matrix of constants is

$$B = \begin{bmatrix} 14.8 \\ 13.5 \\ 18.3 \end{bmatrix}.$$

The product matrix $A^{-1}B$ is $\begin{bmatrix} 0.0075 \\ -0.2672 \\ 14.8000 \end{bmatrix}$, and hence the solution to this system (see Chapter 7) is

$$\begin{aligned}
a &= 0.0075 \\
b &= -0.2672 \\
c &= 14.8000.
\end{aligned}$$

The corresponding quadratic function is

$$MPG_1(t) = 0.0075t^2 - 0.2672t + 14.8$$

and its graph is shown in Figure 8.3.

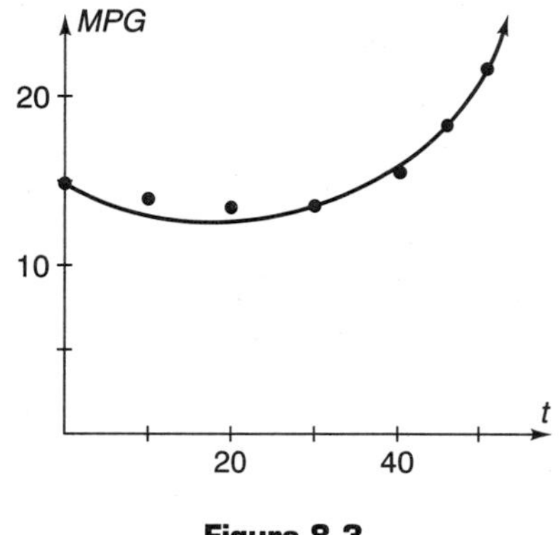

Figure 8.3

Notice that the parabola actually passes through the chosen points (0, 14.8), (30, 13.5), and (46, 18.3). Also notice that the original data point (40, 15.5) is below the parabola; that's because

$$MPG_1(40) = 16.1,$$

which is larger than 15.5.

Let's try one more example. This time use the points (10, 13.9), (20, 13.4), and (30, 13.5). The system of equations is

$$13.9 = 100a + 10b + c$$
$$13.4 = 400a + 20b + c$$
$$13.5 = 900a + 30b + c.$$

The coefficient matrix of the system is

$$A = \begin{bmatrix} 100 & 10 & 1 \\ 400 & 20 & 1 \\ 900 & 30 & 1 \end{bmatrix},$$

and the matrix of constants is

$$B = \begin{bmatrix} 13.9 \\ 13.4 \\ 13.5 \end{bmatrix}.$$

The product matrix $A^{-1}B$ is $\begin{bmatrix} 0.0030 \\ -0.1400 \\ 15.000 \end{bmatrix}$, and the corresponding quadratic function is

$$MPG_2(t) = 0.0030t^2 - 0.1400t + 15.00.$$

Its graph is shown in Figure 8.4.

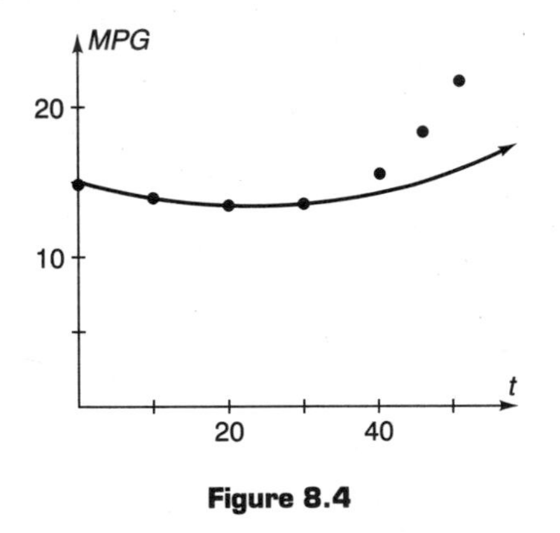

Figure 8.4

Now, we have derived two of the possible thirty-five quadratic functions which could serve as a model for this data. Of course, we must find the best one, and this involves finding the error for each. The errors for parabolic models are computed the same way as the errors for linear models, and the function with smallest error is the best. Table 8.2B shows the actual data values for the relevant years along with the functional values (predicted values) for each of the three derived quadratic functions.

So the best of these two functions is $MPG_1(t)$. Even though this may not (it isn't) be the best function which models this data, we will use it to illustrate the type of information such an equation can provide.

TABLE 8.2B

				Difference for:	
t	Actual	$MPG_1(t)$	$MPG_2(t)$	MPG_1	MPG_2
0	14.8	14.8	15.0	0.0	0.2
10	13.9	12.9	13.9	1.0	0.0
20	13.4	12.5	13.4	0.9	0.0
30	13.5	13.5	13.5	0.0	0.0
40	15.5	16.1	14.2	0.6	1.3
46	18.3	18.4	14.9	0.1	3.4
51	21.7	20.7	15.7	1.0	6.0
			Error	3.6	10.9

EXAMPLE 8.1

In order to determine what the average fuel efficiency will be in 1995, you first need to determine t. The variable t represents the number of years after 1940, so $t = 55$. Next, use your calculator to evaluate the function at 55, that is,

$$MPG_1(55) = 22.8 \text{ mpg.} \quad \blacktriangle$$

EXAMPLE 8.2

Many environmentalists are lobbying for a fuel efficiency of 45 mpg by the year 2000. Suppose that automobile manufacturers do not yield to this pressure, and that the trend shown by our function continues. In what year would fuel efficiency actually reach 45 mpg? To decide this, you need to solve the quadratic equation

$$45 = MPG_1(t),$$
$$45 = 0.0075t^2 - 0.2672t + 14.8.$$

Put this into standard form:

$$0 = 0.0075t^2 - 0.2672t - 30.2,$$

and use the quadratic formula to get

$$t = \frac{-(-0.2672) \pm \sqrt{(-0.2672)^2 - 4(0.0075)(-30.2)}}{2(0.0075)},$$

or $t = -48.1$ or $t = 83.7$. Should $t = -48.1$, then the year would have been

$$1940 - 48 = 1892.$$

Of course, there were very, very few cars then! So the answer is $t = 83.7$, or 84 rounded. The year will be $1940 + 84 = 2024$, and that's twenty-four years after the environmentalists' target year of 2000.

If you use your calculator graph to solve this equation, you would use the trace feature to estimate the first coordinate when the second coordinate is approximately 45. See if you can determine when fuel efficiency will reach 30 mph by using the trace feature. The answer is $t = 66$, or in the year $1940 + 66 = 2006$. ▲

EXAMPLE 8.3

When do you think fuel efficiency was poorest? This involves finding the vertex of the parabola. Remember, when a parabola opens up, the second coordinate of its vertex is its minimum value. The first coordinate is given by

$$t = -\frac{b}{2a} = \frac{-(-0.2672)}{2(0.0075)} = 17.8.$$

This corresponds to the year 1958. The gas mileage was only

$$MPG_1(18) = 12.4 \text{ mpg}.$$

Not too great. Some of you may remember, or have seen pictures of, those huge Cadillacs with the big fins—major contributors to atmospheric CO_2. But, weren't they spectacular!

Figure 8.5 shows the graph of $MPG_1(t)$ with relevant points indicated. ▲

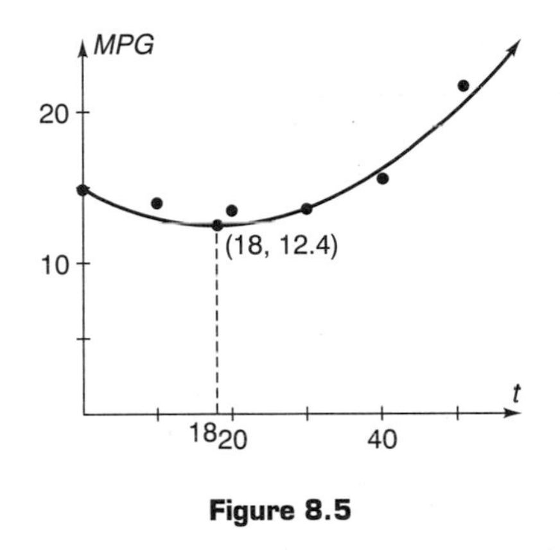

Figure 8.5

It is your task to come up with the best parabola for this data. You should do this in your group after your instructor has assigned years to you. There will be a total of thirty-three remaining equations. Each group should report its best parabola to the class. Also, show its graph, do some predictions for future years, and answer this question: "When was gasoline efficiency at its worst?"

Once the reports are done, you will know the best function for gas mileage; call it $MPG(t)$; save it for future use.

Discuss what social, political, or physical changes might effect the accuracy of this model.

The following is an interesting study on gasoline and carbon dioxide.

1. What is the gas mileage on your car?
2. How many gallons of gasoline does your car burn in one mile?
3. If average fuel efficiency improves each year, cars will burn less gasoline per mile. Use the equation for $MPG(t)$ derived in this section to write the function $GPM(t)$ which defines gallons of gasoline burned each mile in terms of year $1940 + t$.
4. Enter this function $GPM(t)$ into your calculator and graph. Use "trace" to determine its maximum, and explain what this means.
5. Predict the average number of gallons of gasoline which will be burned per mile per car in the year 1998.
6. Each gallon of gasoline burned emits 20 pounds of CO_2. Use your answer to Exercise 2 above to determine the number of pounds of CO_2 your car emits each mile you drive. Estimate the total number of miles you drive each year, and find out how many pounds of CO_2 your car emits in a year.
7. Use your answer to Exercise 5 above to predict the average number of pounds of CO_2 which will be emitted per mile per car in the year 1998.

Use the following information to complete the methanol study below.

i) It takes 75 percent more methanol than gasoline to go one mile.

ii) When burned, one gallon of methanol emits 9.5 pounds of CO_2.

8. Use your answer to Exercise 3 in the above gasoline study to determine the function which defines gallons of methanol which would be burned each mile in terms of year $1940 + t$.

9. Use your answer to the preceding problem to predict the average number of gallons of methanol which would be burned per mile per car in the year 1998.

10. Use your answer to the preceding problem to predict the average number of pounds of CO_2 per mile per car which would be emitted in the year 1998.

Answers to Exercises 5, 7, 9, and 10 will be needed in Chapter 19.

EARTH NOTE Offshore Drilling

It is estimated that from drilling offshore in the Alaskan National Wildlife Refuge, a maximum production of approximately one million barrels of oil per day could be achieved by the year 2002. This level of production could continue through 2010, but then depletion would cause production from this source to decrease to zero by 2020.

Environmental organizations point out that it is technologically possible to manufacture all automobiles with a fuel efficiency rating of 40 mpg without effecting performance. Although it may not be feasible to implement such a change overnight, it seems reasonable that it could be done a few years from now. So, let us suppose that manufacturers are required to increase fuel efficiency so that average fuel efficiency is 40 mpg by the year 2002, and the manufacturers decide to satisy this by a straight line increase beginning this year.

1. Using the *MPG* function, determine the average fuel efficiency this year. Use this information together with the target *MPG* of 40 mpg in 2002 to write a linear equation which defines target fuel efficiency each year from now through 2002.

For Exercises 2–6, assume that each car is driven an average of 14×10^3 miles per year.

2. Use the functions $A(t)$ and $MPG(t)$ derived in Sections 8.1 and 8.2 to write the function $G(t)$ which determines the total

number of gallons of gasoline burned each year under current trends.

3. Use the function $G(t)$ to determine the number of gallons of gasoline burned in the years 2002 and 2010.
4. Suppose that average gas mileage remains constant at 40 mpg from 2002 through 2010. Use this and $A(t)$ to compute the number of gallons of gasoline which would be burned in the years 2002 and 2010.
5. How much gasoline could be saved in the years 2002 and 2010 if 40 mpg were the average gas mileage?
6. One barrel of oil yields 21 gallons of gas (from Bill Skeen, Oklahoma oil man). Write a paragraph comparing the oil savings from the 40 mpg requirement with the amount which could be obtained from drilling offshore in ANWR during its maximum production period.

Source: Kaufman, Donald, and Franz, Cecelia, *Biosphere 2000: Protecting our Global Environment*, HarperCollins College Publishers (New York, 1994).

8.3 AVERAGE YEARLY MILEAGE OF CARS

There's a great line in a great movie called "Repo Man": "The more you drive, the less intelligent you are." If you haven't seen the movie, check it out from your favorite video store. Table 8.3 indicates the average

TABLE 8.3

Year	Avg. Miles Driven per Car per Year ($\times 10^3$)
1940	9.1
1950	9.1
1960	9.5
1970	10.0
1980	8.8
1986	9.3
1991	10.7

Source: *Statistical Abstract of the United States, 1993*, Bureau of the Census.

number of miles each automobile in the United States is driven in the designated year. Take $t = 0$ to be 1940 and plot the corresponding points on a (t, M) coordinate system, where

$$M(t) = \text{average miles driven (in thousands) per car in year } 1940 + t.$$

Figure 8.6 shows these points on the graph. The pattern made by these points is probably unfamiliar to you—it certainly doesn't look like any ONE of the functions listed as possibilities for use as a model in *Earth Algebra.* The key word in the last sentence is ONE. Now, before reading on, look at the figure and think about how you might model this data.

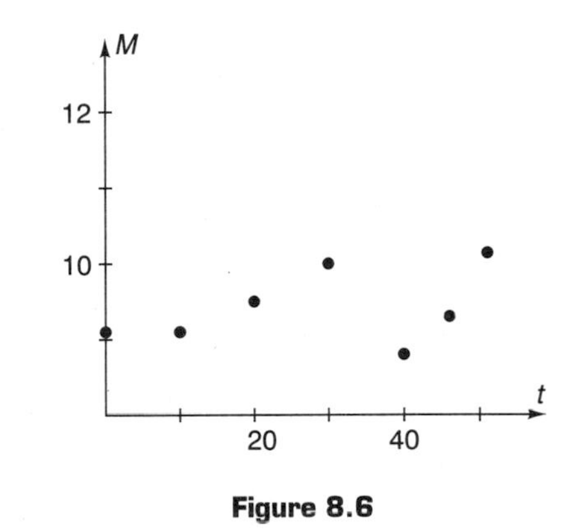

Figure 8.6

What we see (we, meaning the authors) is not just one curve, but two. Maybe some of you saw a couple of lines, one from 1940 to 1970 and the other from 1980 on into the future. The problem with using two lines is that you would end up with a big gap in your overall graph—neither good for looking back into the past, nor describing continuous trends. The data is best modeled with two curves, the first going from 1940 to 1970, and the second from 1970 on into the future. These curves should have a common point in the year 1970 to avoid the discontinuity mentioned above. The data for 1940–1970 possibly could be approximated with a line, but a parabola is a better fit (why?) We also chose a parabola for 1970 and the years following in order to preserve the continuity and closely approximate the points.

This kind of function is called a *piecewise function:* "piecewise" because its definition changes over different time intervals, although you still have only one function. For this model, we have this requirement: the parabola for 1940–1970 must have the point corresponding to 1970 at its right hand endpoint, and the parabola which starts in 1970 and goes on into the future must have the same point for its left hand endpoint. In other words, the first parabola ends at the same point where the second one begins—this avoids the big gap in the graph.

More specifically, you have two sets of data points. The first set is in Table 8.4A; the second set is in Table 8.4B. This is just the data in Table 8.3 broken into two parts, with 1970 common to each.

TABLE 8.4A

Year	Miles per Car
1940	9.1
1950	9.1
1960	9.5
1970	10.0

TABLE 8.4B

Year	Miles per Car
1970	10.0
1980	8.8
1986	9.3
1991	10.7

The task is to find the best parabola which models the data in Table 8.4A, and then to find the best one which models the data in Table 8.4B. The equation for any parabola for 1940–1970 must pass through the point (30, 10.0), and the same goes for any parabola for 1970–future.

After all this is done, you have two equations, but only one function which models the data in Table 8.3. This is a piecewise function. Here is an example. One quadratic equation for 1940–1970 is

$$M_1(t) = .0005t^2 + .0250t + 8.8000$$

(subscript 1 for first interval),

which is derived using points (10, 9.1), (20, 9.5), and (30, 10.0). It is good for approximating data from 1940–1970 only. A quadratic equation for 1970–1991 is

$$M_2(t) = 0.0154t^2 - 1.2154t + 32.5875$$

which is derived using points (30,10.0), (46, 9.3), and (51, 10.7). But this one can only be used for the years 1970 on. Finally, the ONE function is the piecewise function $M(t)$ defined by

$$M(t) = \begin{cases} 0.0005t^2 + 0.025t + 8.8 & \text{if } 0 \leq t \leq 30, \\ 0.0154t^2 - 1.2154t + 32.5875 & \text{if } t \geq 30, \end{cases}$$

and its graph is shown in Figure 8.7.

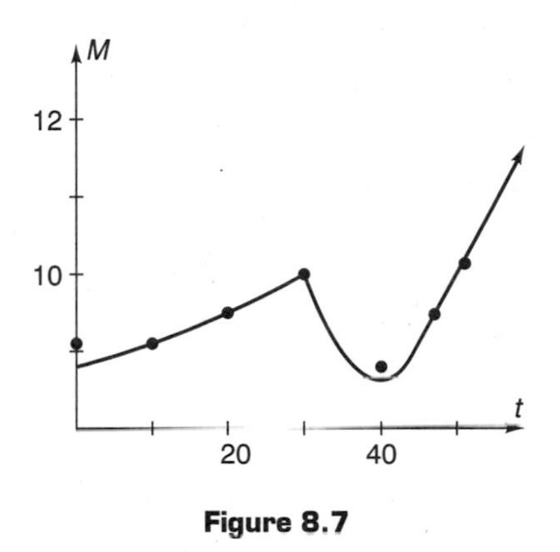

Figure 8.7

Write all possible quadratic equations which model the data in Table 8.4A—you must always use as one point (30, 10.0). When you finish writing all such equations (there are only three), find the error using only points corresponding to 1940–1970. The equation with the smallest error is the best one for the time interval 1940–1970.

Next, repeat the same exercise for years from 1970–1991. Again, you must always use (30, 10.0) as one point. Find the one with the smallest error; this is the one to use for years 1970 on.

Use your model to answer the following questions:

1. Estimate the number of miles each car drove in 1953.
2. Estimate the number of miles each car will drive in 1998.
3. When will the average number of miles driven be 12,000?
4. When was the average number of miles driven 9700?

5. Discuss any social, political, or physical changes which might affect the accuracy of this model. What might be reasonable limitations on the domain and range of this function?
6. Suppose you had constructed this model using only data available through 1970. How would this affect the accuracy of your predictions?

EARTH NOTE Average Highway Speed

Did you know that even in 1988, the average speed on Los Angeles freeways was only 35 mph?

Source: Kaufman, Donald, and Franz, Cecelia, *Biosphere 2000: Protecting our Global Environment,* HarperCollins College Publishers (New York, 1994).

CHAPTER 9

Modeling Carbon Dioxide Emission from Power Consumption in the United States

In this chapter energy consumption data in the United States are provided (*Statistical Abstracts of the United States*). Fuels are coal, petroleum, and natural gas, and quantities are measured in quadrillion BTU (quadrillion means 10^{15}), abbreviated "quads." Each set of data is to be modeled by the groups, and finally total energy consumption will be determined.

EARTH NOTE Peanut Butter

One BTU (British thermal unit) is the amount of energy required to raise the temperature of one pound of water one degree Fahrenheit at or near 39.2 degrees F; this is approximately the equivalent of one match tip. 1000 BTU of energy is equivalent to eight-tenths of a peanut butter and jelly sandwich; 1,000,000 BTU corresponds to eight gallons of gasoline (or 800 peanut butter and jelly sandwiches); one quadrillion (10^{15}) BTU is equivalent to twenty-six hours of world energy use in 1989.

Source: (*Environmental Quality: 22nd Annual Report,* The Council on Environmental Quality,) U.S. Government Printing Office, 1992.

9.1 PREDICTING COAL CONSUMPTION

Here are the figures for coal consumption in the U.S. (Table 9.1):

TABLE 9.1

Year	Coal Consumption (quads)
1970	12.28
1975	12.67
1980	15.43
1985	17.46
1990	19.11

Source: *Statistical Abstract of the United States, 1993*, Bureau of the Census.

Figure 9.1 shows these data plotted on a (t, CC) coordinate system where $CC(t)$ is coal consumption (in quads) in year $1970 + t$.

Since all of the information starts with 1970, we let $t = 0$ correspond to 1970.

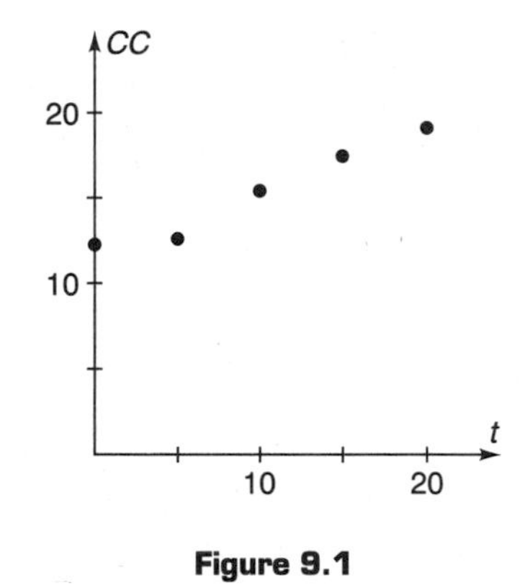

Figure 9.1

The earlier points fit nicely into a parabola, but the later three points appear to be nearly collinear. Therefore, we suggest a piecewise function as a model: a parabola for years 1970–1980, and a line for years 1980 future. Remember, all possible lines for the later years must begin at the common point, in 1980, in order to preserve continuity.

It is a task for the class to determine the best model for coal consumption. Your final model will be only ONE function, although it is defined by two different equations.

Use your model to obtain the following information.

1. Predict the coal consumption for the year 1997.
2. What was coal consumption in 1976?
3. When will coal consumption be 25 quads?
4. Discuss social, political, or physical changes which might affect the accuracy of this model.

9.2 PREDICTING PETROLEUM CONSUMPTION

Data for consumption of petroleum are in Table 9.2.

TABLE 9.2

Year	Petroleum Consumption (quads)
1970	29.48
1975	32.85
1980	34.20
1985	30.93
1990	33.58

Source: *Statistical Abstract of the United States, 1993,* Bureau of the Census.

The corresponding points are plotted on a (t, PC) coordinate system (Figure 9.2) with $t = 0$ for 1970, and $PC(t) =$ petroleum consumption (in quads) in year $1970 + t$.

Once again, do you see two different patterns? You can use a piecewise curve again, but this time use a linear function for the first (years 1970–1980), and a quadratic for the second (1980–future). All lines for the early years must end at the point corresponding to 1980.

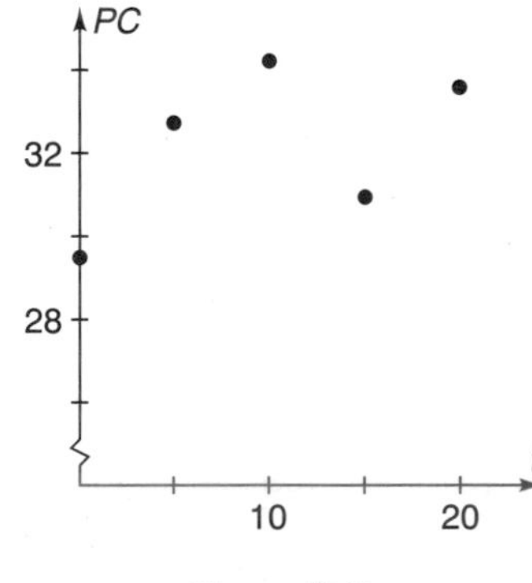

Figure 9.2

The class should now derive the best model for petroleum consumption. Use your model to derive the following information.

1. Predict petroleum consumption in year 2001.
2. What was petroleum consumption in 1973?
3. When was petroleum consumption 33 quads?
4. When is consumption equal to 40 quads?
5. Discuss social, political, or physical changes which might affect the accuracy of this model.

9.3 PREDICTING NATURAL GAS CONSUMPTION

Data for consumption of natural gas are in Table 9.3.

TABLE 9.3

Year	Natural Gas Consumption (quads)
1970	21.78
1975	20.03
1980	20.44
1985	17.83
1990	19.27

Source: *Statistical Abstract of the United States, 1993*, Bureau of the Census.

The points corresponding to the data in Table 9.3 are plotted in Figure 9.3 on a (t, NGC) coordinate system, where $NGC(t) =$ natural gas consumption (in quads) in year $1970 + t$.

This suggests (we hope) another piecewise curve, in this case two parabolas with 1980 serving as the common year.

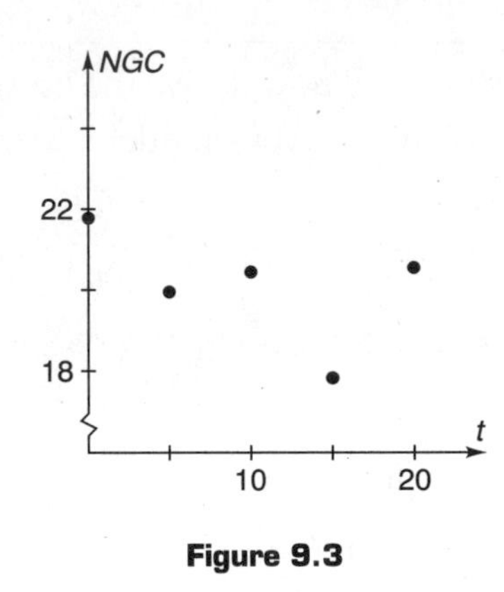

Figure 9.3

The class should derive the best model for natural gas consumption. Use your model to determine the following.

1. Predict natural gas consumption in 2010.
2. What was consumption in 1976?
3. In what year(s) will natural gas consumption once again reach the 1970 level?
4. When was consumption least? What was the amount?
5. Discuss social, political, or physical changes which could affect the accuracy of this model.

9.4 PREDICTING TOTAL ENERGY CONSUMPTION

The patterns for total energy consumption are more complicated since they involve more factors; included in the total consumption figures are other sources of power, such as solar, nuclear, wood burning, and geothermal. It seems that for the three energy sources we studied, patterns dramatically changed in 1980, so for our total consumption model, we concentrate only on data obtained for years from 1980 through 1990.

TABLE 9.4

Year	Total Energy Consumption (quads)
1980	76.0
1982	70.8
1985	74.0
1987	76.8
1990	81.3

Source: *Statistical Abstract of the United States, 1993,* Bureau of the Census.

Plot all the data points on a (t, TEC) coordinate system, where $TEC(t)$ is total energy consumption (in quads) in year $1970 + t$. We think you will agree that a parabola appears to be a good model. The class should now determine the best parabola to use to model the data.

Use your model to determine the following.

1. In what year was energy consumption the least?
2. Predict energy consumption for the year you plan to graduate from college.
3. In what year will energy consumption reach 120 quadrillion BTU?
4. Discuss social, political, or physical changes that might effect the accuracy of your model.

Save all models derived in this chapter. You'll need them later.

EARTH NOTE Appliance Efficiency

More efficient household appliances can dramatically reduce the electricity consumed, and this savings in electricity in turn decreases the emission of carbon dioxide. The chart below provides some estimates for electricity consumed by conventional appliances, more efficient models, and models with highest efficiency. Units are kilowatt hours per year.

Appliance	Conventional	Improved Efficiency	Best Efficiency
Refrigerator	1500	1100	750
Central air	3600	2900	1800
Electric water heater	4000	2900	1800
Electric range	800	750	700

It is also estimated that electricity generated from coal produces 2.27 pounds of carbon dioxide per kilowatt hour. Estimate how much carbon dioxide emission your household is responsible for each year from use of these appliances.

Source: Kaufman, Donald,and Franz, Cecelia, *Biosphere 2000: Protecting our Global Environment,* HarperCollins College Publishers (New York, 1994).

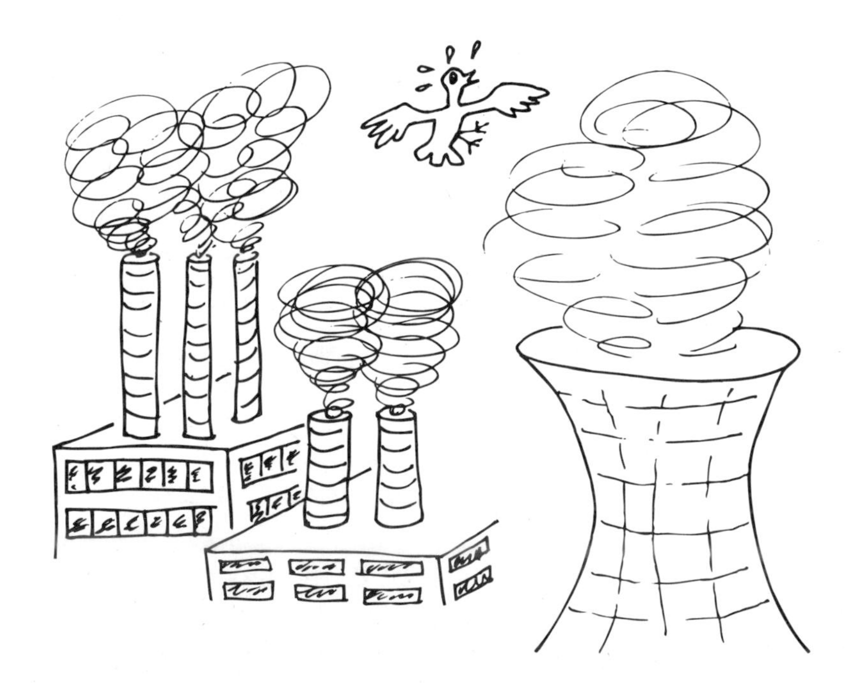

Exponential and Logarithmic Functions

10.1 EXPONENTIAL FUNCTIONS

The general form we use for an *exponential function* is

$$f(x) = a(b^x),$$

where a and b are constant, and x is the variable. The variable x is called the *exponent,* and the constant b is called the *base*. Restrictions on b are required: b must be positive and different from 1. (Note that if $b = 1$, then this function would be constant, and the restriction of b to positive values allows the domain of an exponential function to be all real numbers x.) Regardless of the value of x, $b^x > 0$, and hence the range of an exponential function will be all $y > 0$ if $a > 0$, and all $y < 0$ if $a < 0$.

Exponential functions are easily graphed on your calculator; here are some examples.

Note: Chapter 10 is a prerequisite for Chapters 11 and 14.

EXAMPLE 10.1

Graph $f(x) = 2^x$ on your calculator. (Here, $a = 1$ and $b = 2$.)

Enter this function into your calculator, and graph. You should see the graph as shown in Figure 10.1. The "range" for this graph is set as follows:

$$x \text{ min} = -2;\ x \text{ max} = 2;\ y \text{ min} = 0;\ y \text{ max} = 10.$$

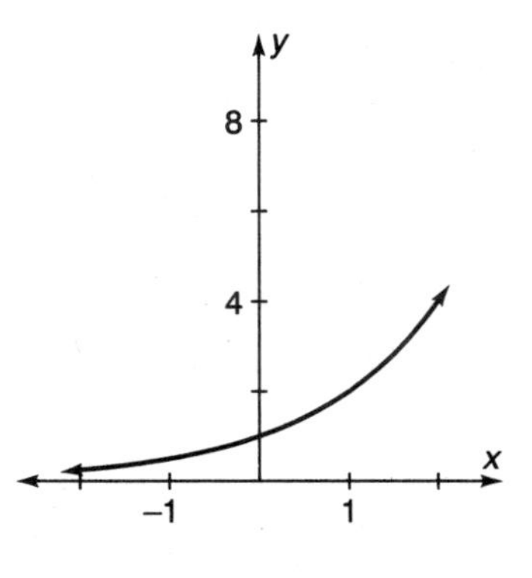

Figure 10.1

You should note that there are no x-intercepts; the y-intercept is (0, 1) which means $2^0 = 1$, and the entire graph is above the x-axis which means that $2^x > 0$ for all x. ▲

EXAMPLE 10.2

Leave the function $f(x) = 2^x$ in your calculator, and enter this second function

$$g(x) = 10^x.$$

(Here $a = 1$ and $b = 10$.) Now press **graph,** and you should see the graph of both exponential functions on your screen (Figure 10.2). ▲

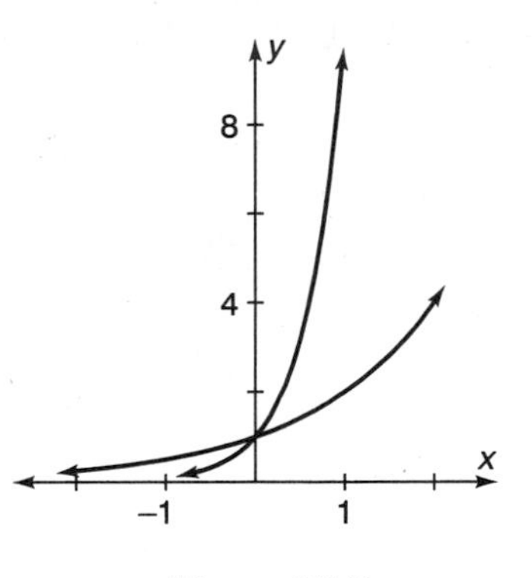

Figure 10.2

These graphs look basically the same except for the rate of increase. Note that the graph of $y = 10^x$ is much steeper than the graph of $y = 2^x$ because $10 > 2$. In general, the larger the base, the steeper the graph, which means that the function increases really fast when the base is big.

EXAMPLE 10.3

Graph $h(x) = (0.5)^x$ on your calculator. Now $b = 0.5 < 1$ and the graph should appear as in Figure 10.3. This graph is decreasing, whereas the graph of $y = 2^x$ is increasing. If the base b is smaller than 1, the exponential function $y = b^x$ decreases, and if the base b is larger than 1, $y = b^x$ increases. Also, note from Figure 10.3 that $(0.5)^x > 0$. There are no x-intercepts and the y-intercept is (0, 1). ▲

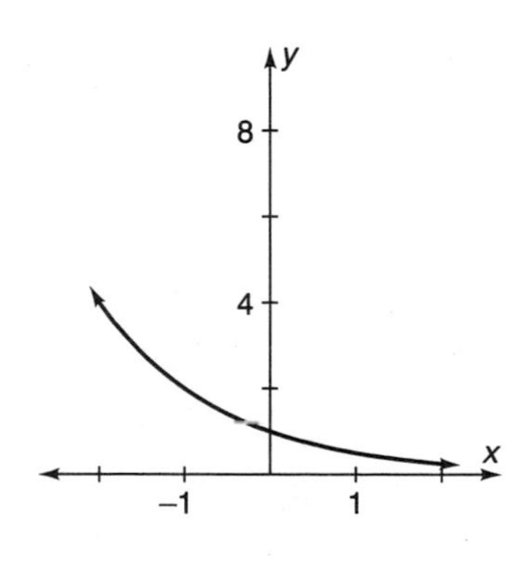

Figure 10.3

EXAMPLE 10.4

Enter a second function, $y = .12^x$, and graph both functions. You should see something like Figure 10.4. For these two functions, $y = .12^x$ decreases more rapidly. ▲

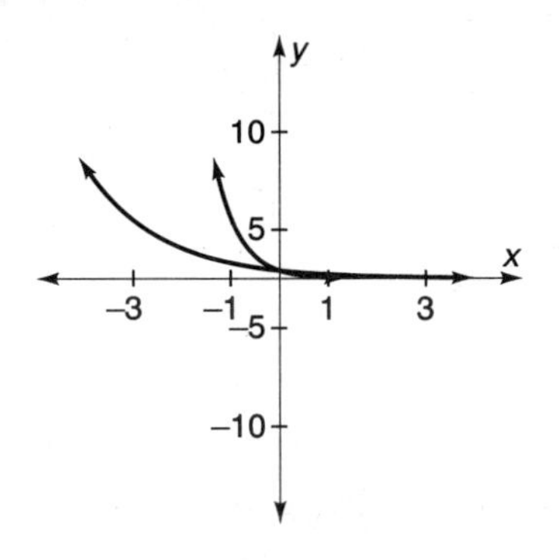

Figure 10.4

In general, the graph of the exponential function $f(x) = b^x$ looks like one of the two shown in Figure 10.5.

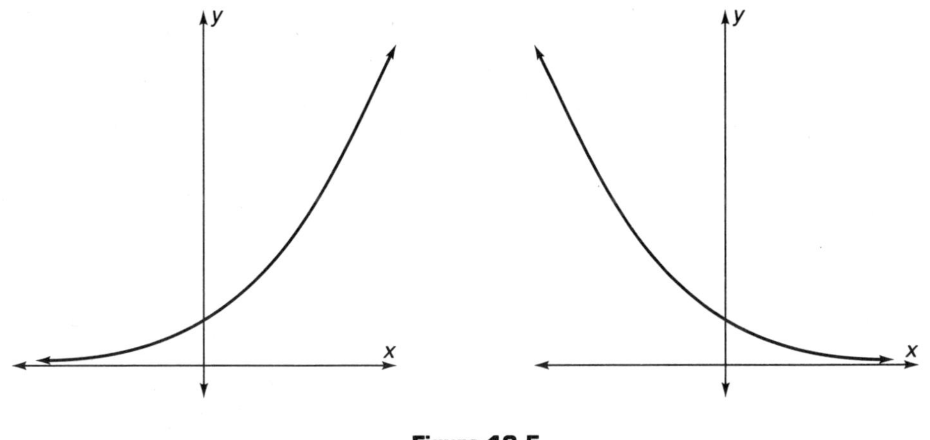

Figure 10.5

We list some properties which are true for all exponential functions of the form $y = b^x$.

1. There are no x-intercepts.
2. The y-intercept is (0, 1) since $b^0 = 1$.
3. The entire graph is above the x-axis since $b^x > 0$ for all x.
4. If $0 < b < 1$, the function decreases; the smaller the base b, the greater the rate of decrease.
5. If $b > 1$, the function increases; the larger the base b, the greater the rate of increase.

Properties 6–10 are the ones which are commonly known as the "laws of exponents."

6. $b^p \times b^q = b^{p+q}$.
7. $\dfrac{b^p}{b^q} = b^{p-q}$.
8. $(b^p)^q = b^{pq}$.
9. $b^{-p} = \dfrac{1}{b^p}$.
10. $\dfrac{1}{b^{-p}} = b^p$.

EXAMPLE 10.5

Graph $y = 0.53(1.07)^x$ on your calculator, and determine the y-intercept.

Enter this function into your calculator and graph (Figure 10.6). In this example,

$$a = 0.53,$$

which makes the y-intercept (0, 0.53). ▲

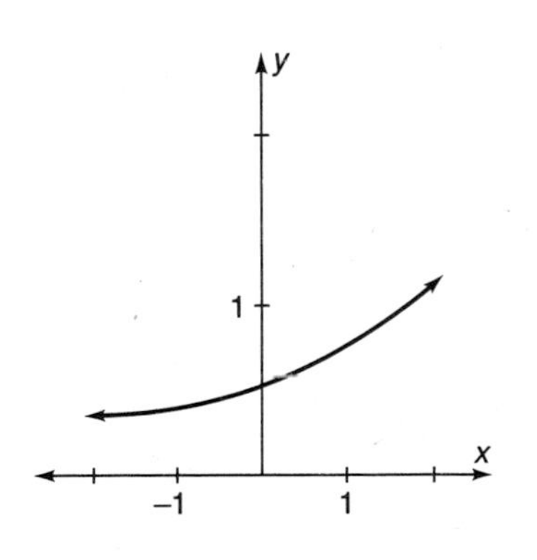

Figure 10.6

EXAMPLE 10.6

A very special number in mathematics is known as "e" with approximate value 2.71828. This number e occurs quite naturally in a variety of studies, in particular, in physics, biology, and even economics. The exponential function

$$y = e^x$$

is known as the *natural exponential function,* its base is

$$b = e,$$

and its graph is shown in Figure 10.7. The notation "e" is for a famous mathematician, Leonhard Euler (1707–1783). ▲

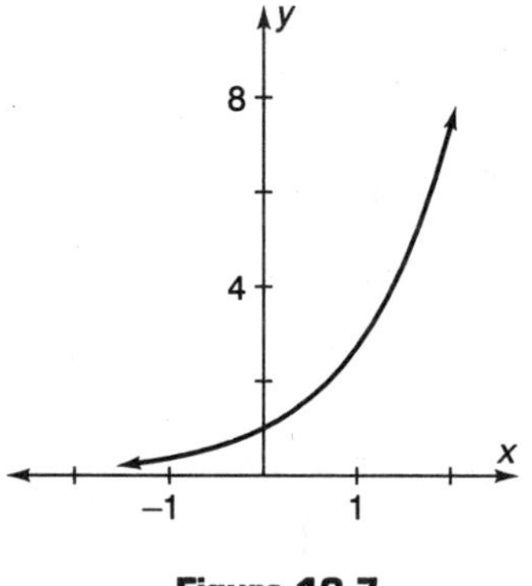

Figure 10.7

10.1 THINGS TO DO

Use your calculator to evalute each of the following.

1. $10^{.259}$
2. $(1.72)^{5.61}$
3. e^2
4. $3e^{2.1}$
5. $7.31(5.6)^{9.74}$

Graph each of the functions in Exercises 6–10 on your calculator and determine the y-intercept.

6. $f(x) = 6^x$

7. $g(x) = -3^x$

8. $h(x) = (1.13)^x$

9. $k(x) = 9.6(4.3)^x$

10. $l(x) = -2.4e^x$

10.2 INVERSE FUNCTIONS

Some functions have related functions which are called their *inverses*. The inverse of a function "undoes" what the function does. If f denotes a function which has an inverse, then its inverse function is denoted by f^{-1}. (This does *not* mean $\frac{1}{f}$.)

In general,

$$\text{if } f(x) = y, \text{ then } f^{-1}(y) = x.$$

This is the defining property of an *inverse function*. Note that the domain of f is the range of f^{-1} and the range of f is the domain of f^{-1}. Also, the inverse of the inverse of a function is the function itself; that is, $(f^{-1})^{-1} = f$. If we form both composites of a function and its inverse, we get

$$f^{-1}(f(x)) = x \quad \text{and} \quad f(f^{-1}(y)) = y.$$

The defining property above also shows us how to find the inverse for a particular function: if $f(x)$ is defined by the equation

$$y = f(x),$$

we can determine the equation which defines the inverse f^{-1} if we can solve for x in terms of y to get

$$x = f^{-1}(y).$$

Two examples illustrate the technique used to determine the inverse of a linear function.

EXAMPLE 10.7

Determine the inverse of $f(x) = 4x + 7$.

To find the inverse of this function, replace $f(x)$ by y to get

$$y = 4x + 7.$$

Then solve for x.

$$y - 7 = 4x$$

$$x = \frac{y - 7}{4},$$

so $f^{-1}(y) = \dfrac{y - 7}{4}$.

We check the two composites.

$$f^{-1}(f(x)) = \frac{(4x + 7) - 7}{4} = x.$$

And similarly,

$$f(f^{-1}(y)) = 4\left(\frac{y - 7}{7}\right) + 7 = y. \quad \blacktriangle$$

EXAMPLE 10.8

Determine the inverse of $f(x) = 1.37x + 132.5$.

Replace $f(x)$ by y:

$$y = 1.37x + 132.5.$$

Solve for x:

$$1.37x = y - 132.5$$

$$x = \frac{y - 132.5}{1.37}.$$

Hence

$$f^{-1}(y) = \frac{y - 132.5}{1.37}.$$

This time we check only one composite, $f(f^{-1}(y))$:

$$f(f^{-1}(y)) = 1.37\left(\frac{y - 132.5}{1.37}\right) + 132.5 = y.$$

Similarly, $f^{-1}(f(x)) = x$. ▲

Not all functions have inverses. If solving for x in terms of y produced two or more x-values, then the "inverse" would not satisfy the definition of function, so the inverse would not exist. See Example 10.9.

EXAMPLE 10.9

Consider the function $f(x) = x^2$. Attempting to find the inverse, replace $f(x)$ by y to get $y = x^2$. Solving for x gives $x = \pm\sqrt{y}$. If, for example, $y = 25$, then there are two values for x, 5 and -5, and the "inverse" is not a function. ▲

All strictly increasing or strictly decreasing functions have inverses because no two different x-values can produce the same y-value. In particular, all nonconstant linear functions have inverses.

10.2 THINGS TO DO

For Exercises 1–5, determine the inverse of the given linear function, then form both composites to show that $f^{-1}(f(x)) = x$ and that $f(f^{-1}(y)) = y$.

1. $f(x) = 2x - 1$

2. $F(x) = 4 - 3x$

3. $g(x) = \frac{x}{5} + 20$

4. $h(x) = 7.314x - 2.001$

5. $f(x) = -5.4x$

The following functions do not have inverses. Explain why.

6. $f(x) = 1 - x^2$

7. $g(x) = x^4$

10.3 LOGARITHMIC FUNCTIONS

All exponential functions $f(x) = b^x$ ($b > 0$, $b \neq 1$) have inverses because they are either strictly increasing ($b > 1$) or strictly decreasing ($0 < b < 1$). The inverse of the exponential function $f(x) = b^x$ is called the logarithmic function with base b. It is denoted $f^{-1}(x) = \log_b x$. Since the domain of the exponential function equals all real numbers, this is the range of the logarithmic function. Similarly, the domain of the logarithmic function is all positive real numbers since this is the range of the exponential function.

By definition, if $y = f(x) = b^x$ then $x = f^{-1}(y) = \log_b y$. Therefore we see that $y = b^x$ if $x = \log_b y$. Similarly, if $y = \log_b x$ then $b^y = x$. Some examples should help your understanding.

EXAMPLE 10.10

Determine $\log_2 8$.

The base here is 2 so

$$\log_2 8 = y \quad \text{if} \quad 2^y = 8.$$

What power do you need to take 2 to in order to get 8? The answer is, of course, 3, since $2^3 = 8$. So

$$\log_2 8 = 3. \quad \blacktriangle$$

EXAMPLE 10.11

Determine $\log_{10} 100$.

Here $\log_{10} 100 = y$ if $10^y = 100$. The answer is 2 because $10^2 = 100$, so

$$\log_{10} 100 = 2.$$

In this example, if we use the functional notation $f(x) = \log_{10} x$, then

$$f(100) = 2. \quad \blacktriangle$$

EXAMPLE 10.12

Determine $\log_{10} 0.1$.

You have to do a little more work here. You must find the exponent in the equation

$$10^y = 0.1.$$

The number 0.1 is the same as $\frac{1}{10}$, which is the same as 10^{-1}, so you have the equation

$$10^y = 10^{-1},$$

so $y = -1$; that is, $\log_{10} 0.1 = -1$. $\blacktriangle$

We now list the properties of logarithmic functions $y = \log_b x$. These will be needed later when we solve both exponential and logarithmic equations.

1. There are no y-intercepts regardless of the base b.
2. The x-intercept is $(1, 0)$; $\log_b 1 = 0$ since $b^0 = 1$.
3. The entire graph is to the right of the y-axis because the domain consists of only positive numbers.
4. $\log_b (MN) = \log_b M + \log_b N$.
5. $\log_b \frac{M}{N} = \log_b M - \log_b N$.
6. $\log_b (M^p) = p \log_b M$.
7. $\log_b b = 1$.
8. $\log_b 1 = 0$.
9. $\log_b b^x = x$.
10. $b^{\log_b x} = x$.

These properties can be derived from corresponding properties of exponents.

We will now examine the graphs of logarithmic functions with base 10 and base e.

EXAMPLE 10.13

Graph $y = \log_{10} x$.

The function $\log_{10} x$ is called the *common logarithmic* function, and is the inverse of the function 10^x. It is usually abbreviated as $\log x$. Your calculator should have a key labeled **LOG,** and so to determine the graph, just enter the function $y = \log x$ and press **graph.** Figure 10.8 shows the graph with

$$x \text{ min} = 0,\ x \text{ max} = 10,\ y \text{ min} = -2,\ y \text{ max} = 2.$$ ▲

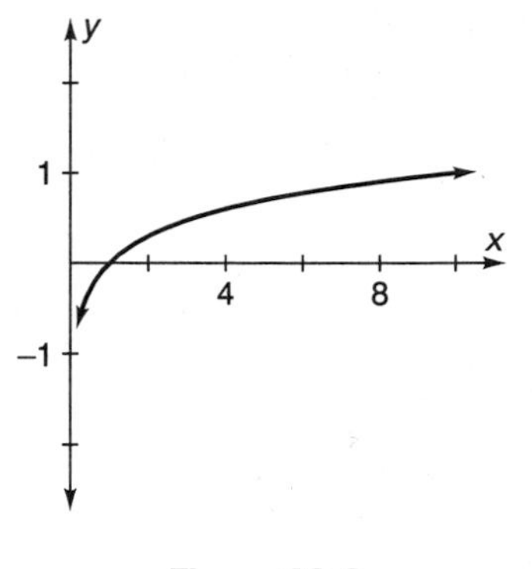

Figure 10.8

The function $\log_e x$ is called the *natural logarithmic function.* It is the inverse of the function e^x and is written $y = \ln x$. We are going to concentrate on natural logarithms since they are the most useful in *Earth Algebra.* The graph of $y = \ln x$, which looks very much like $y = \log x$, is shown in Figure 10.9.

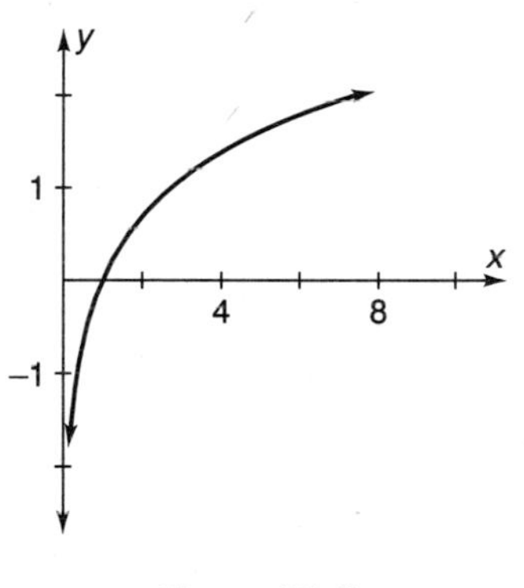

Figure 10.9

The domain of $y = \ln x$ consists of all $x > 0$, and the range is all real numbers y. The graph doesn't go to the left of the y-axis; it increases from $-\infty$ to $+\infty$, but it increases very, very slowly. Its x-intercept is

(1, 0), because $e^0 = 1$ and so $\ln 1 = 0$; it has no y-intercept. The function $y = \ln x$ is easy to evaluate on your calculator. For example, to find ln 3.1 press the **ln** key, then 3.1, then enter:

$$\ln 3.1 = 1.1314.$$

Note that $\ln e = 1$, since $\ln e = \log_e e$ and $e^1 = e$.

The standard form for the logarithmic function which will be used in *Earth Algebra* is

$$y = a + b \ln x,$$

where a and b are constants. We now look at some graphs (on the calculator) of logarithmic functions of this type for different values of a and b.

EXAMPLE 10.14

Graph $y = 3 + 2 \ln x$.

Figure 10.10 shows this graph. Using the trace feature on the calculator, we estimate the x-intercept to be (.22, 0). There is no y-intercept. ▲

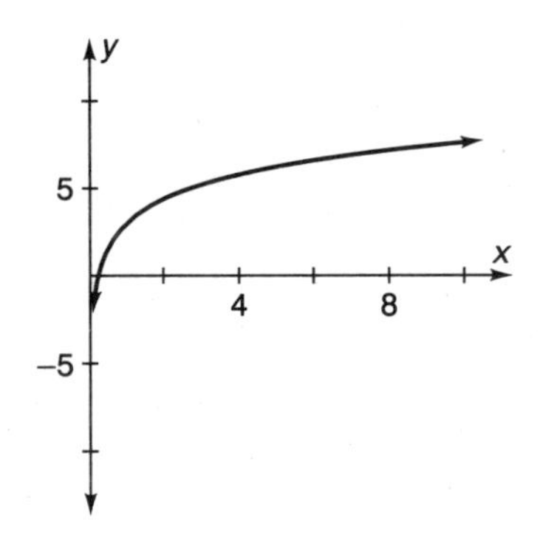

Figure 10.10

EXAMPLE 10.15

Graph $y = 2.2 \ln x - 1.7$.

Figure 10.11 shows this graph. The x-intercept is (1.97, 0) rounded, and again, there is no y-intercept. ▲

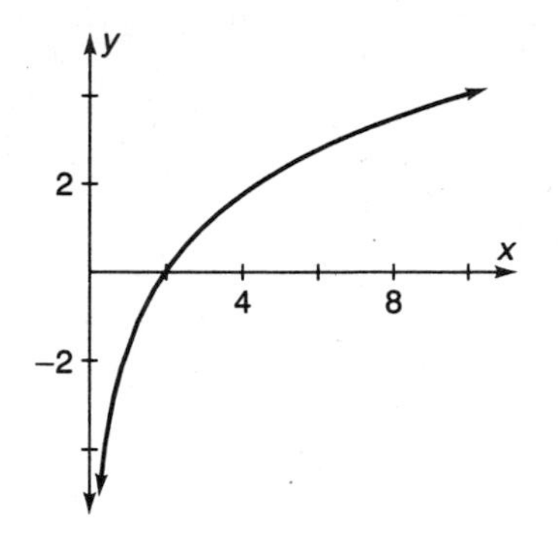

Figure 10.11

EXAMPLE 10.16

Graph $y = 2.3 - 4.5 \ln x$.

Figure 10.12 shows the graph. The x-intercept is estimated to be (1.67, 0). ▲

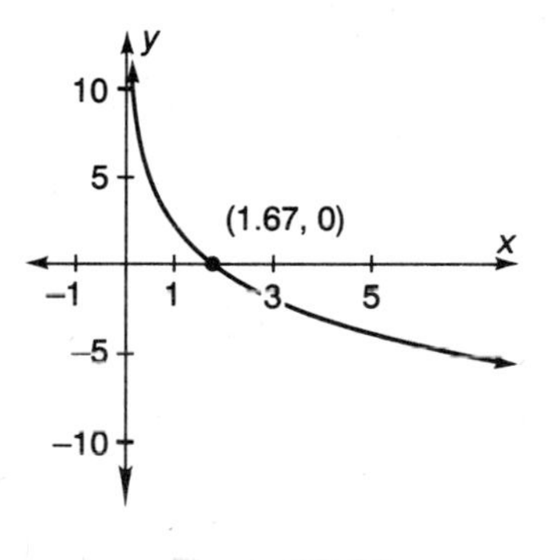

Figure 10.12

10.3 THINGS TO DO

For Exercises 1–7 determine the logarithms; use your calculator for Exercises 4–7.

1. $\log_3 81$
2. $\log_{10} .0001$
3. $\log_{10} 10000$

4. log 17

5. log .023

6. ln 2

7. ln .03

For Exercises 8–12, graph the given equation; estimate the x-intercept for each.

8. $y = 3 + 5 \log x$

9. $y = 5 - 2 \log x$

10. $y = 2 \ln x$

11. $y = 1 - 3.2 \ln x$

12. $y = 5 + 2.3 \ln x$

10.4 LOGARITHMIC AND EXPONENTIAL EQUATIONS

Consider Example 10.15 in Section 10.3. We estimate the x-intercept of this graph by using the trace feature on the calculator to get $x = 1.97$. To find this intercept algebraically, it is necessary to solve the equation

$$0 = 2.5 \ln x - 1.7.$$

First solve for ln x:

$$1.7 = 2.5 \ln x$$
$$\ln x = 0.68.$$

Therefore $e^{\ln x} = e^{0.68}$, and recalling from Property 10 of logarithms that $e^{\ln x} = x$, we have

$$x = e^{0.68} = 1.9739, \text{ rounded to four places.}$$

EXAMPLE 10.17

Let $f(x) = 1.6 \ln x + 4$, and algebraically determine the x-intercept of its graph.

Solve the equation

$$0 = 1.6 \ln x + 4$$

$$-4 = 1.6 \ln x$$

$$\ln x = -2.5$$

Therefore

$$e^{\ln x} = e^{-2.5}$$

$$x = e^{-2.5} = 0.0821. \quad \blacktriangle$$

EXAMPLE 10.18

Let $f(x) = 6 + 8.1 \ln x$. Determine x if $f(x) = 4$.

Solution A. Estimate the x-value when $f(x) = 4$ using the **trace** on your calculator.

To do this, graph $y = 6 + 8.1 \ln x$ and the horizontal line $y = 4$ on your calculator (Figure 10.13).

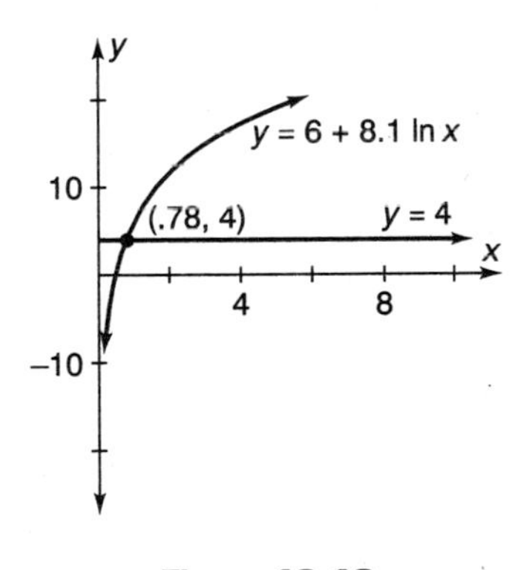

Figure 10.13

The x-value we want to estimate is the x-coordinate of the point of intersection of these graphs. Use **trace** to find

$$x = .78.$$

Solution B. Solve for x algebraically when $f(x) = 4$.

We need to solve the equation

$$4 = 6 + 8.1 \ln x$$
$$-2 = 8.1 \ln x$$
$$\ln x = -0.2469,$$

so $x = e^{-0.2469} = 0.7812$.

Note that the point of intersection of the graphs is (0.7812, 4). ▲

EXAMPLE 10.19

Solve the exponential equation

$$e^{2x+3} = 7.$$

Let $f(x) = e^{2x+3}$.

Solution A. Estimate the x-value when $f(x) = 7$. Graph $y = e^{2x+3}$ and the horizontal line $y = 7$ on your calculator (Figure 10.14). The x-value we want to estimate is the x-coordinate of the point of intersection of these graphs. Use **trace** to find

$$x = -0.53.$$

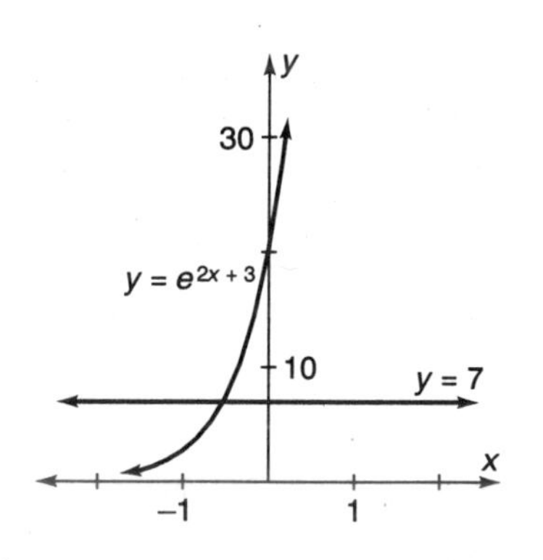

Figure 10.14

Solution B. Solve for x algebraically when $f(x) = 7$.

$$7 = e^{2x+3}.$$

Take the natural logarithm of each side to get

$$\ln 7 = \ln (e^{2x+3}).$$

Recall from Property 9 that $\ln (e^z) = z$, so we get

$$\ln 7 = 2x + 3.$$

Next evaluate ln 7 on the calculator and solve for x:

$$\begin{aligned} 1.9459 &= 2x + 3 \\ 2x &= -1.0541 \\ x &= -0.5270. \quad \blacktriangle \end{aligned}$$

EXAMPLE 10.20

Solve the exponential equation

$$2(1.2^x) = 5.$$

First divide by 2 to isolate the "exponent part."

$$(1.2^x) = 2.5.$$

Next take the natural logarithm of each side:

$$\ln (1.2)^x = \ln (2.5);$$

then recall Property 9 of logarithms which says that $\ln b^x = x \ln b$. This gives

$$x \ln (1.2) = \ln (2.5);$$

Now we can solve for x:

$$x = \frac{\ln (2.5)}{\ln (1.2)} = 5.0257. \quad \blacktriangle$$

10.4 THINGS TO DO

A. Graph each function on your calculator and use the graph to solve the given equation.
B. Solve each equation algebraically.

1. $f(t) = 1.2 + 3.7 \ln t$; solve $1.2 + 3.7 \ln t = 9$.

2. $y = 500 + 225 \ln x$; solve $500 + 225 \ln x = 100$.

3. $H(s) = .002 - .035 \ln s$; solve $.002 - .035 \ln s = .04$.

4. $g(x) = e^{x+1}$; solve $e^{x+1} = 2$.

5. $r(t) = e^{8t-4}$; solve $e^{8t-4} = 1$.

6. $p(t) = 1.7(3^t)$; solve $1.7(3^t) = 4.5$.

10.5 INVERSES OF EXPONENTIAL AND LOGARITHMIC FUNCTIONS

The technique for finding the inverse of an exponential or logarithmic function is the same as that for a linear function, but a little more difficult.

EXAMPLE 10.21

Determine the inverse of $f(x) = 2e^x$.

Replace $f(x)$ by y to get

$$y = 2e^x.$$

Note that you need to solve for x, which is the exponent, and property 9 of logarithmic functions enables you to do this. First divide by 2 to get

$$\frac{y}{2} = e^x,$$

then take the logarithm of each side: to get

$$\ln \frac{y}{2} = \ln (e^x) = x \ln e.$$

Recall from Section 10.2 that $\ln e = 1$.

Since $\ln e = 1$

$$\ln \frac{y}{2} = x,$$

or

$$f^{-1}(y) = \ln \frac{y}{2}. \quad \blacktriangle$$

EXAMPLE 10.22

Determine the inverse of $f(x) = e^{2x+3}$. Now

$$y = e^{2x+3}$$
$$\ln y = \ln e^{2x+3}$$
$$\ln y = 2x + 3.$$

Solve for x:

$$\ln y - 3 = 2x$$
$$x = \frac{\ln y - 3}{2}$$

So $f^{-1}(y) = \dfrac{\ln y - 3}{2}$.

To illustrate the relationship between f and its inverse, we observe

$$f^{-1}(f(x)) = f^{-1}(e^{2x+3}) = \frac{\ln (e^{2x+3}) - 3}{2}$$
$$= \frac{(2x + 3)\ln e - 3}{2}$$
$$= \frac{2x + 3 - 3}{2}$$
$$= \frac{2x}{2} = x.$$

Without computation, do you know what $f^{-1}(f(2))$ is? $\blacktriangle$

EXAMPLE 10.23

Determine the inverse of $f(x) = 2^x$.

While the definition of a logarithmic function says that this is $\log_2 y$, there is no $\log_2$ key on most calculators, so for convenience we illustrate how the inverse of $f(x) = 2^x$ can be expressed in terms of natural logarithms. Set

$$\begin{aligned} y &= 2^x \\ \ln y &= \ln 2^x \\ \ln y &= x \ln 2 = x(.693), \end{aligned}$$

and

$$x = \frac{\ln y}{.693},$$

so

$$f^{-1}(y) = \frac{\ln y}{.693}. \quad \blacktriangle$$

EXAMPLE 10.24

Determine the inverse of $f(x) = 2.35(1.1^x)$.

Set $y = 2.35(1.1^x)$ and solve for x. First you must divide by 2.35 to get

$$\frac{y}{2.35} = 1.1^x;$$

then take logarithms:

$$\begin{aligned} \ln\left(\frac{y}{2.35}\right) &= \ln (1.1^x) \\ &= x \ln (1.1) \\ &= x(0.95). \end{aligned}$$

This gives

$$x = \frac{1}{.095} \ln \left(\frac{y}{2.35} \right)$$

so

$$f^{-1}(y) = \frac{1}{.095} \ln \left(\frac{y}{2.35} \right). \quad \blacktriangle$$

The following examples illustrate techniques for finding inverses of natural logarithmic functions. You will need to recall these properties of logarithms:

$$e^{\ln x} = x \quad \text{and} \quad \ln e^x = x.$$

EXAMPLE 10.25

Determine the inverse of $f(x) = 2 + 3 \ln x$.

Replace $f(x)$ by y to get

$$y = 2 + 3 \ln x.$$

Solve for $\ln x$:

$$\ln x = \frac{y - 2}{3};$$

Thus

$$e^{\ln x} = e^{\frac{y-2}{3}},$$

so

$$x = e^{\frac{y-2}{3}},$$

and hence

$$f^{-1}(y) = e^{\frac{y-2}{3}}.$$

We check composites both ways.

$$f(f^{-1}(y)) = 2 + 3\ln\left(e^{\frac{y-2}{3}}\right)$$
$$= 2 + 3\,\frac{(y-2)}{3}$$
$$= y,$$

and

$$f^{-1}(f(x)) = e^{\frac{(2+3\ln x)-2}{3}} = e^{\ln x} = x,$$

just as it should be. ▲

EXAMPLE 10.26

Determine the inverse of $g(x) = 3.7 + 5\ln(2x)$.

Set

$$y = 3.7 + 5\ln(2x).$$

and solve for $\ln(2x)$

$$\ln(2x) = \frac{y-3.7}{5}$$

so

$$e^{\ln(2x)} = e^{\frac{y-3.7}{5}},$$
$$2x = e^{\frac{y-3.7}{5}}$$
$$x = \frac{1}{2}e^{\frac{y-3.7}{5}}.$$

Thus

$$g^{-1}(y) = \frac{1}{2}e^{\frac{y-3.7}{5}}.$$

Now check the composites:

$$\begin{aligned} g(g^{-1}(y)) &= 3.7 + 5 \ln \left(2\left(\frac{1}{2}\right)e^{\frac{y-3.7}{5}}\right) \\ &= 3.7 + 5 \ln e^{\frac{y-3.7}{5}} \\ &= 3.7 + 5\left(\frac{y - 3.7}{5}\right) = y; \end{aligned}$$

and

$$\begin{aligned} g^{-1}(g(x)) &= \frac{1}{2}e^{\frac{(3.7+5 \ln (2x))-3.7}{5}} \\ &= \frac{1}{2}e^{\ln(2x)} \\ &= \frac{1}{2}(2x) \\ &= x. \quad \blacktriangle \end{aligned}$$

10.5 THINGS TO DO

For Exercises 1–5, determine the inverse of the given exponential function by using natural logarithms, form both composites to show that $f^{-1}(f(x)) = x$ and $f(f^{-1}(y)) = y$, and compute the indicated functional values.

1. $g(x) = 3^x$. Compute $g(2)$ and $g^{-1}(9)$.
2. $h(x) = 2.1\ (7.01)^x$. Compute $h(4.3)$ and $h^{-1}(9095.085205)$.
3. $f(x) = (6.03)^x$. Compute $f(2.5)$ and $f^{-1}(2.5)$.
4. $g(x) = -3(1.2)^x$. Compute $g(1)$ and $g^{-1}(-3.6)$.
5. $F(x) = 1 + 5^x$. Compute $F(0)$ and $F^{-1}(2)$.

For Exercises 6–10, determine the inverse of the given logarithmic function, form both composites to show that $f^{-1}(f(x)) = x$ *and that* $f(f^{-1}(y)) = y$, *and compute the indicated functional values.*

6. $f(x) = 2.7 + 1.3 \ln x$; compute $f(2)$ and $f^{-1}(3.601)$.

7. $g(x) = 5 - 3 \ln x$; compute $g(7)$ and $g^{-1}(-.8377)$.

8. $h(x) = 1.5 \ln (3x) - 9$; compute $h(8)$ and $h^{-1}(-4.2329)$.

9. $L(x) = \ln (5 - x)$; compute $L(0)$ and $L^{-1}(1.6094)$.

10. $U(x) = .874 \ln (2x)$; compute $U(-3)$.

CHAPTER 11

Modeling Carbon Dioxide Emission Resulting from Deforestation

11.1 PREDICTING DEFORESTATION DUE TO LOGGING

Governments of developing countries are ready and willing to export forest products, particularly timber, in order to obtain currency from foreign countries. Large timber companies from North America and

Note: The data provided in this chapter is projected from information obtained in Gaia: *An Atlas of Planet Management*. This information was also used to compile Tables 11.1, 11.2, and 11.3 in this chapter.

Japan are happy to collect profits from imported rain forest woods. Not only does this contribute to global warming, but it also depletes supply for local consumption, in particular, for heating and cooking. Essentially, poor forestry practice is harmful for all parties involved.

The information provided below (Table 11.1) is the estimated number of hectares (ha) of rain forest lost to logging by year.

TABLE 11.1 Rain Forest Cut for Logging

Year	ha × 10^6
1960	.44
1970	1.19
1980	2.20
1988	3.90
1990	4.29

Since we will model other rain forest data with logarithmic functions (which have only positive numbers in their domain), we want the first year, 1960, to have positive t-value so groups should model these data with $t = 0$ in 1940. We think you will agree that this looks parabolic. Your instructor will assign years to each group. When your model is complete, name your best function $L(t)$, and preserve it.

Use your model to determine the following information.

1. How many hectares will be cut for logging in the year 2005?
2. There are 11.6×10^6 ha of ancient forests in the U.S. In what year will the area of rain forest cut for logging alone equal this number?

EARTH NOTE Diversity in Rain Forests

Rain forests, which cover only 7 percent of the surface of the earth, are incredibly rich in diversity of species; it is estimated that anywhere from 50 percent to 90 percent of the earth's species are found in the tropical rain forests, many of which are not found anywhere else in the world. Thus deforestation has a serious effect on the biological diversity of the planet. According to the United States Academy of Sciences, ten square kilometers may contain 1500 species of flowering plants, 750 tree species, 400 bird species, 150 butterfly species, 100 reptile species, 60 amphibian species, and an unknown number of insect species.

Source: Kaufman, Donald, and Franz, Cecelia, *Biosphere 2000: Protecting our Global Environment,* HarperCollins College Publishers (New York, 1994).

11.2 PREDICTING DEFORESTATION DUE TO CATTLE GRAZING

The destruction of a major portion of the rain forests in Central and South America has been to provide grazing lands for cattle. Cattle ranching has been encouraged by local governments, and most of the resulting beef went to the United States for use at fast food restaurants and for pet foods. Due to inefficiency, cattle ranching has decreased recently; indeed, most rain forest soil does not provide good pasture, and has limited use. It takes approximately two and one-half acres to feed one cow, and in addition to that, erosion and compactification render the land unusable after only a few years.

The data provided below (Table 11.2) is the estimated amount of rain forest cleared for cattle ranching in the indicated year.

TABLE 11.2 Rain Forest Cut for Cattle Grazing

Year	ha × 10^6
1960	0.30
1970	0.95
1980	1.36
1990	1.66

The graph (Figure 11.1) shows these points plotted on a (t, CG) coordinate system with $t = 0$ in 1940.

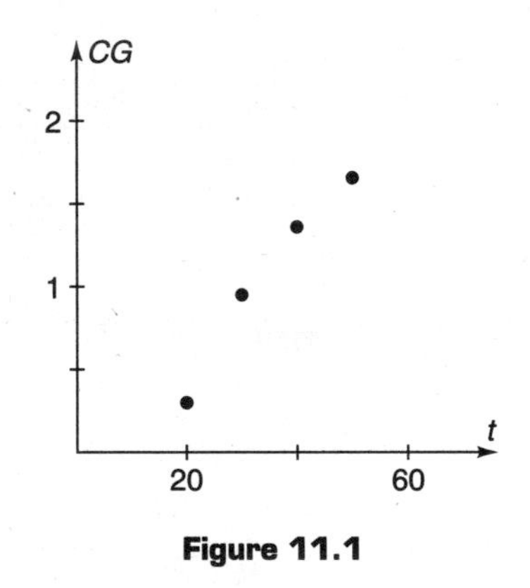

Figure 11.1

We need to discuss the type of equation to use to model this data. Note that the number of hectares cut is increasing from 1950 through 1990, but the rate of increase is slowing from 1950 through 1990 (slopes between consecutive data points are .065, .041, and .030). Hence a linear model would not be good. Your next guess might be a parabolic model. This would be OK except for one problem: the vertex would occur in some year in the future, which would indicate that the amount of land cleared for cattle would reach a maximum and then start decreasing after that year. Without any information to support such an event, this could give very inaccurate predictions. Therefore, it would be nice to have an equation which more accurately reflects this decreasing rate of increase, but with no vertex in the future. What curve do you know that behaves like that? The one that we will use is a logarithmic curve, with the general form

$$y = k(\ln x) + c,$$

where k and c are constants. Adapting our variables and using

t = number of years after 1940,

$CG(t)$ = number of hectares ($\times 10^6$) cleared for cattle grazing in year $1940 + t$, we have

$$CG(t) = k(\ln t) + c.$$

To use this type of equation as a model, we need to substitute two points from the data provided in order to determine the constants (two constants, two points). Here's an example.

Use the years 1960 and 1980; the points are (20, 0.30) and (40, 1.36). Then substitute into the general form to get the two equations

$$0.30 = k(\ln 20) + c$$
$$1.36 = k(\ln 40) + c,$$

or

$$0.30 = 2.9957k + c$$
$$1.36 = 3.6889k + c.$$

This is a system of 2 equations in 2 unknowns with 2×2 coefficient matrix

$$A = \begin{bmatrix} 2.9957 & 1 \\ 3.6889 & 1 \end{bmatrix},$$

and 2×1 constant matrix

$$B = \begin{bmatrix} 0.30 \\ 1.36 \end{bmatrix}.$$

Enter these matrices into your calculator, and determine the product $A^{-1}B$ to get the solution

$$A^{-1}B = \begin{bmatrix} 1.5293 \\ -4.2812 \end{bmatrix},$$

so

$$k = 1.5293$$
$$c = -4.2812.$$

The equation is

$$CG(t) = 1.5293 (\ln t) - 4.2812.$$

Differences and error are computed as usual. First evaluate the function at the t-values:

$$t = 20,\ CG(20) = 1.5293 (\ln 20) - 4.2812 = 0.30$$
$$t = 30,\ CG(30) = 1.5293 (\ln 30) - 4.2812 = 0.92$$
$$t = 40,\ CG(40) = 1.5293 (\ln 40) - 4.2812 = 1.36$$
$$t = 48,\ CG(48) = 1.5293 (\ln 48) - 4.2812 = 1.70$$

Next compute the differences and add them to obtain the error.

Year	Predicted	Actual	Difference
1960	0.30	0.30	.00
1970	0.92	0.95	.03
1980	1.36	1.36	.00
1990	1.70	1.66	.04
			Error = .07

And finally, the graph is shown in Figure 11.2.

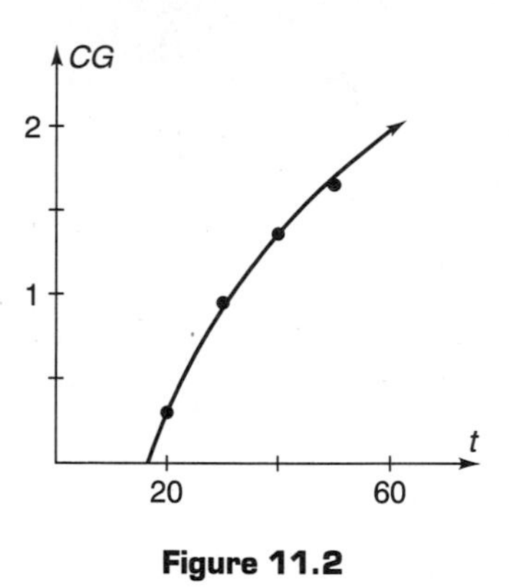

Figure 11.2

Although this may not be the best function to model our data, we use it to illustrate some predictions with a logarithmic function. For example, in the year 2001($t = 61$) we predict that the amount of rain forest to be cleared for cattle grazing will be

$$\begin{aligned} CG(61) &= 1.5293 \ln(61) - 4.2812 \\ &= 2.01, \end{aligned}$$

so in 2001, 2.01 million hectares will be cleared.

Next, we predict the year in which 2.5 million hectares will be cleared. This involves finding a solution to the equation

$$2.5 = 1.5293 \ln t - 4.2812.$$

We solve this algebraically by first solving for $\ln t$:

$$\begin{aligned} 1.5293 \ln t &= 6.7812 \\ \ln t &= \frac{6.7812}{1.5293} = 4.4342, \end{aligned}$$

then $e^{\ln t} = e^{4.4342}$, and $t = 84$ years after 1940, or in the year 2024. Also, this solution can be estimated by using the calculator to find a point on the graph with second coordinate near 2.5. Remember to use "trace." You should see a point near (84.3, 2.50); this means $t = 84$ (rounded) when $CG = 2.50$. Hence the solution is approximately $t = 84$, and the year would be $1940 + 84 = 2024$. See Figure 11.3.

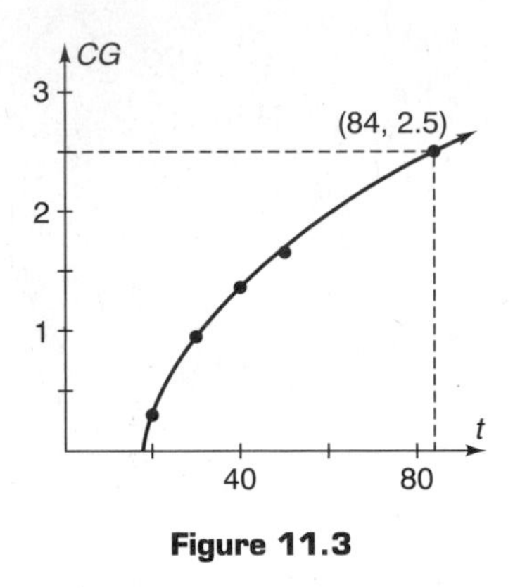

Figure 11.3

Now you have seen one example of writing a logarithmic equation to match the data provided. It's almost as easy as writing a linear equation, only now you are using logarithms.

The groups should now find the remaining logarithmic equations for this data and make their reports. There are only six equations total, and we've already done one of them. Call the best function $CG(t)$, and save it. Use $CG(t)$ to determine this information:

1. In what year will the amount of rain forest cut for cattle ranching equal 3 million hectares? Solve algebraically and check your answer by estimation on the calculator.
2. How many hectares will be cleared for cattle ranching in the year 2005?
3. Discuss social, political, or physical changes which might affect the accuracy of this model.

11.3 PREDICTING DEFORESTATION DUE TO AGRICULTURE AND DEVELOPMENT

Although agriculture has been encouraged locally, rain forest soils are so lacking in nutrients necessary for crop growth, it is proving to be too inefficient to pursue. Farmers would abandon their small farms after the few existing nutrients were used up, leaving the land fallow and useless. It takes up to 30 years for the land to regain enough nutrients to farm again, and then only for a short time.

Development plans in most tropical countries include roads, dams, and resettlement of peasant populations. Many of these projects have been, at best, minimally successful, and the damage to the rain forests has proved to be more detrimental than beneficial.

The following data (Table 11.3) are the estimated amount of rain forest destroyed for agricultural and developmental projects by year. This table provides the information for the third and final factor which we will use in our analysis of deforestation.

TABLE 11.3 Rain Forest Destroyed for Agriculture and Development

Year	ha $\times 10^6$
1960	2.21
1970	3.79
1980	4.92
1990	5.94

Groups should now model this information, and report. We *strongly* suggest a logarithmic model here, for the samc rcasons as before. Use $t = 0$ in 1940, and name your best equation $AD(t)$. Use $AD(t)$ to determine the following: in what year will the amount cut for agriculture and development be five times the amount destroyed for cattle in 1995? Solve algebraically, and check your result by estimation with the calculator.

Discuss future changes which might affect the accuracy of your model.

You have now completed your model for the three major factors involved in deforestation. In the next chapter, we build models which define total emission of carbon from each of the three sources: automobiles, energy consumption, and deforestation.

Total Carbon Dioxide Emission Functions

Now that you've got the model for each factor from the three sources of carbon dioxide emission, each group should concentrate on one of the sources, and write a function which gives total CO_2 or carbon emission from that source as a function of time t. We break this chapter down into three sections. Each group should read the section appropriate for its assigned (or chosen) source.

12.1 PASSENGER CARS

The three factors involved here are: total number of passenger cars, $A(t)$; average fuel efficiency, $MPG(t)$; and average miles per car per year, $M(t)$. The conversion necessary to obtain CO_2 emission is this: each gallon of gasoline burned emits 20 lbs of carbon dioxide (on the average).

In order to determine the function which defines CO_2 emission from automobiles, you must first determine the total number of gallons of gasoline actually burned. You should use only the part of the piecewise function $M(t)$ for which $t \geq 30$ since we will use the total emission function for future predictions. Name this function $CO_2A(t)$, and clearly explain your variables. Note that the domain of $CO_2A(t)$ is $t \geq 30$. Save this important function.

Enter $CO_2A(t)$ in your calculator so that you can easily evaluate and graph it. Maybe you can enter the equation for each factor and then combine them. Before you graph this function, you'll need to set the calculator ranges so that you can see the relevant parts of the graph. For example, if you want to see trends through the year 2050, then you should set x min at 30, and x max at $110(2050 - 110 = 1940)$ or more. To estimate your vertical range, calculate $CO_2A(30)$ and set y min a little less than this; then calculate $CO_2A(110)$ and set y max a little more than this number.

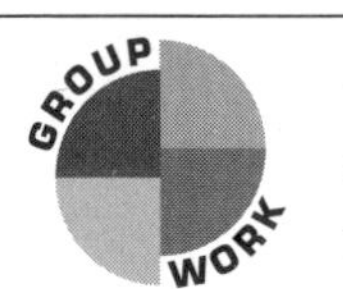

Suppose you wanted to know the year in which CO_2 emission from passenger cars in the United States will reach the astronomical figure of 2000×10^9 lbs. This involves solving the equation.

$$CO_2A(t) = 2000,$$

which is quite difficult. But you can estimate the solution by locating the point on the graph with vertical coordinate approximately 2000.

Or, maybe you'd need to know CO_2 emission from all these cars when you're sixty-four. To do this, you need to know how old you are now.

See what other interesting things you can predict or otherwise glean from your function. For example, do you see any high or low points on your graph? If so, explain these. Include all of this in your group report for the class.

12.2 POWER CONSUMPTION

You have modeled the amount of power consumed from three sources: coal, natural gas, and petroleum. These equations predict how many quadrillion BTU of each fuel will be consumed in any particular year. One reason for wanting to predict how much energy will be consumed in the United States is so that we can predict how much carbon dioxide will be emitted from these sources. The three fuels whose consumption you have modeled are called carbon-bearing fuels because when burned each of them emits carbon in the form of carbon dioxide; we must determine how much carbon is due to each of these sources. You have the model for each source: coal consumption, $CC(t)$; petroleum consumption, $PC(t)$; natural gas consumption, $NGC(t)$.

To determine the carbon emission from each of these, you need to know how much carbon is emitted per quad for each fuel. Table 12.1 below gives the necessary information. In order to keep our numbers

TABLE 12.1 Carbon Content of Fuels

Fuel	Gigatons per Quad
Coal	.02500
Natural Gas	.01454
Petroleum	.02045

Source: "Changing by Degrees: Steps to Reduce Greenhouse Gases," U.S. Congress, Office of Technology Assessment, 1986.

relatively small, we choose "gigatons of carbon" to measure emission; one gigaton is 10^{12} kilograms. Recall that one quad is 10^{15} BTU. (See Appendix B for all units.)

We wish to know how much carbon is emitted from coal, natural gas, and petroleum. For example, to obtain total carbon emitted from the burning of coal in year $1970 + t$, multiply $CC(t)$ by the factor .025.

Power groups should now determine carbon emission from coal, natural gas, and petroleum, and then combine all three sources to obtain a total carbon emission function for energy consumption. (Remember that these are piecewise functions; since we will use these to predict future events, you should use the latter definition, i.e., $t \geq 10$.) Name it $TCE(t)$; this defines total carbon emission from energy consumption in gigatons in year $1970 + t, t \geq 10$.

Enter $TCE(t)$ into your calculator so that you can easily evaluate and graph it. Maybe you can enter the equation for each factor and then combine them. Before you graph this function, you'll need to set the calculator ranges so that you can see the relevant parts of the graph. For example, if you want to see trends through the year 2050, you should set x min at 10 and x max at 80 ($2050 - 80 = 1970$) or more. To estimate your vertical range, set y min a little less than the minimum value of the function; then calculate $TCE(80)$ and set y max a little more than this number.

Suppose you need to know the year in which carbon emission from energy consumption will reach 10 gigatons. This involves solving the equation $TCE(t) = 10$ for t, which is easy with the calculator. Estimate the solution by locating the point on the graph with second coordinate approximately equal to 10. What is your estimate for the first coordinate?

Or, maybe you would like to know carbon emission from energy consumption when you're sixty-four. To do this, you need to know how old you are now.

See what other interesting things you can predict or otherwise glean from your function. For example, do you see any high or low points on your graph? If so, explain these. Include all of this in your group report for the class.

12.3 DEFORESTATION

The three major reasons for deforestation are logging, cattle grazing, and agriculture and development. You have derived models for each of these factors; retrieve them. To obtain carbon emission, the key conversion is: one hectare of destroyed rain forest emits one half of a metric ton of carbon. (Here we only consider carbon emitted from forests cut.)

Rain forest groups should now combine the three models for the factors to determine a carbon emission function for deforestation. Name it $CDF(t)$; this defines total carbon emission in metric tons from deforestation in year $1940 + t$.

Enter $CDF(t)$ into your calculator so that you can easily evaluate and graph it. Maybe you can enter the equation for each factor and then combine them. Before you graph this function, you'll need to set the calculator ranges so that you can see the relevant parts of the graph. For example, if you want to see trends through the year 2050, then you should set x min at 0, and x max at $110(2050 - 110 = 1940)$ or more.

Estimation of the vertical range is a little tricky because there are logarithms in your function. This means that $CDF(0)$ is undefined (the domain of the natural logarithmic function consists only of positive numbers), so you should calculate CDF at some small positive number, say $CDF(5)$ and set y min to this number or a little less. Then calculate $CDF(110)$ and set y max a little more than this.

Suppose you wanted to know the year in which carbon emission from deforestation in the United States will reach 10×10^6 tons. This involves solving the equation $CDF(t) = 10$, which is quite difficult. But you can estimate the solution by locating the point on the graph with second coordinate approximately 10. What is your estimation?

Or, maybe you need to know carbon emission from deforestation when you're sixty-four. To do this, you need to know how old you are now.

See what other interesting things you can predict or otherwise glean from your function. For example, do you see any high or low points on your graph? If so, explain these. Include all of this in your group report for the class.

PART III

Accumulation of Carbon Dioxide

Introduction

In Part III, our primary objective is to study the accumulation of carbon emission due to human activity and how this affects the atmospheric concentration of CO_2. In order to carry out this study, in Chapter 14 we first derive a new model for long-term atmospheric CO_2 concentration using a more sophisticated function, namely an exponential function. This easily enables us to determine the growth rate of atmospheric CO_2. Then we use the most recent information on annual carbon emission and its growth rate to predict future carbon emission and its contribution to atmospheric accumulation.

In Chapter 15, we approximate the models developed in earlier chapters with exponential models, and use these new models to estimate atmospheric CO_2 accumulation due to each of the three sources studied.

The mathematical concepts needed to conduct these studies are linear and exponential functions and geometric series. Chapter 13 covers geometric series.

A Reminder. Coefficients of functions and equations in all environmental chapters should be rounded as follows: linear, two places; quadratic, exponential, and logarithmic, four places. Sometimes we deviate from this practice but not without warning.

Some discrepancy may occur in answers if round-off is done during calculation as opposed to only rounding the final result. In all situations the final answer in all the environmental chapters should be rounded to the same number of places as the original data.

Geometric Series

13.1 NOTATION, DEFINITIONS, AND SUMS

A finite *geometric progression* is a finite sequence of numbers, each of which is obtained from its predecessor by multiplication by a fixed constant. The general form for a finite geometric progression is

$$a, ar, ar^2, ar^3, \ldots, ar^n,$$

where a can be any number, the fixed constant is r, and n can be any positive integer. The number r is called the *common ratio.* An example of a particular geometric progression is

$$3, 6, 12, 24, 48.$$

Here $a = 3$, $r = 2$, and $n = 4$.

A finite geometric series, or simply *geometric series,* is

$$a + ar + ar^2 + \cdots + ar^n;$$

that is, the sum of the terms of a finite geometric progression. The *sum*

$$S = a + ar + ar^2 + \cdots + ar^n$$

Note: Chapter 13 is a prerequisite for Chapters 14 and 15.

of a geometric series is easily determined by the formula

$$S = \frac{a(r^{n+1} - 1)}{r - 1}.$$

(See Appendix D for a derivation of this formula.) The following examples illustrate how this formula works.

EXAMPLE 13.1

The geometric series

$$3 + 6 + 12 + 24 + 48$$

corresponds to the progression given as your earlier example. Recall,

$$a = 3$$
$$r = 2$$
$$n = 4.$$

Substitute into the formula to get

$$S = \frac{3(2^5 - 1)}{2 - 1} = 93.$$

Thus

$$93 = 3 + 6 + 12 + 24 + 48. \quad \blacktriangle$$

EXAMPLE 13.2

Consider the geometric series

$$5 + 15 + 45 + 135 + 405 + 1215 + 3645 + 10{,}395.$$

First, determine

$$a = 5$$
$$r = 3$$
$$n = 7.$$

So

$$S = \frac{5(3^8 - 1)}{3 - 1} = 16{,}400. \quad \blacktriangle$$

EXAMPLE 13.3

Consider the series

$$5.31 + 5.31(.26) + 5.31(.26)^2 + 5.31(.26)^3.$$

In this form it is easy to see that

$$a = 5.31$$
$$r = .26$$
$$n = 3,$$

so

$$S = \frac{5.31((.26)^4 - 1)}{.26 - 1} = 7.1429. \quad \blacktriangle$$

This last example (13.3) is given in the form

$$a + ar + ar^2 + \cdots + ar^n$$

because the geometric series you will encounter in *Earth Algebra* will come to you in this form. Note that this makes it very easy to determine a, r, and n.

You can also write your own geometric series if you know the numbers a, r, and n. The next example illustrates this point.

EXAMPLE 13.4

Let $a = 7$, $r = 4$, and $n = 6$.

The corresponding series is

$$7 + 7(4) + 7(4)^2 + 7(4)^3 + 7(4)^4 + 7(4)^5 + 7(4)^6,$$

or multiplied out,

$$7 + 28 + 112 + 448 + 1792 + 7168 + 28{,}672.$$

Its sum is

$$S = \frac{7(4^7 - 1)}{4 - 1} = 38{,}227. \quad \blacktriangle$$

13.1 THINGS TO DO

For Exercises 1–5, determine the numbers a, r, and n, and the sum of each geometric series.

1. $1 + 6 + 36 + 216 + 1296 + 7776 + 46{,}656 + 279{,}936$

2. $8 + 4 + 2 + 1 + .5 + .25 + .125 + .0625$

3. $71.2 + 71.2(4.3) + 71.2(4.3)^2 + 71.2(4.3)^3 + \cdots + 71.2(4.3)^{12}$

4. $1.6 + 1.6(.431) + 1.6(.431)^2 + 1.6(.431)^3 + 71.2(.431)^4$

5. $.5 + .5(.1) + .5(.1)^2 + \cdots + .5(.1)^{793}$

For Exercises 6–10, write the geometric series corresponding to the given values of a, r, and n, and find its sum.

6. $a = 3, \quad r = 8, \quad n = 5$

7. $a = 4.1, \quad r = 9.6, \quad n = 7$

8. $a = 2.16, \quad r = 1.7, \quad n = 52$

9. $a = .1, \quad r = .01, \quad n = 10$

10. $a = 2, \quad r = 3, \quad n = 1$

Carbon Dioxide Concentration Revisited

14.1 LONG-TERM CARBON DIOXIDE CONCENTRATION

When we first began modeling, we modeled atmospheric concentration of carbon dioxide with a linear function, which wasn't a bad first try. Let's now look at more long-term data and try a different model which will better reflect these trends. The measurements are made at the Mauna Loa Observatory in Hawaii. (See Table 14.1)

We plot the corresponding points on a (t, CO_2) coordinate system (Figure 14.1) where

$$CO_2(t) = \text{carbon dioxide concentration in ppm in year } 1900 + t.$$

TABLE 14.1

Year	Average CO_2 Concentration in ppm
1900	280.0
1950	310.0
1975	331.0
1980	338.5
1990	354.0

Source: World Resources, 1992–1993, World Resources Institute, 1993.

(Note: for this model $t = 0$ in 1900, so t = number of years after 1900.) This graph is shown in Figure 14.1.

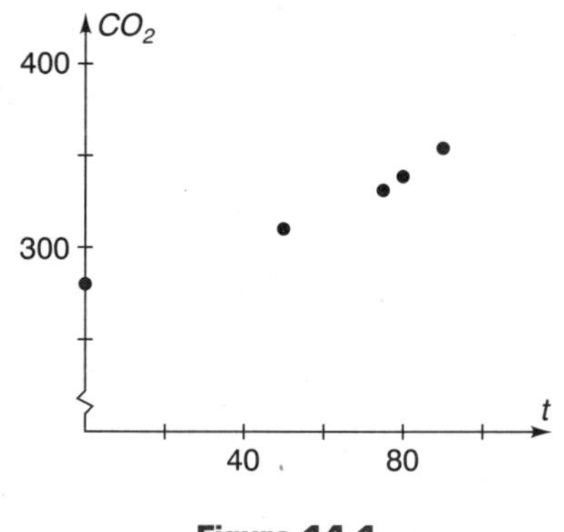

Figure 14.1

In order to decide on the best type of curve to use for this model, we compute the slopes of the line segments which join adjacent points. The slope between the first two points, (0, 280) and (50, 310), is

$$m = \frac{310 - 280}{50 - 0} = .6$$

and the slope between (50, 310) and (75, 331) is

$$m = \frac{331 - 310}{25} = .84.$$

The next two slopes, in order, are 1.5 and 1.55. These slopes are increasing, so either a parabolic or exponential model will provide a better fit than a linear one. Since there is no indication of a vertex in our data, we use an exponential model.

Recall the general form for an exponential function

$$f(x) = a(b^x),$$

where a and b are constant (Chapter 10). Adapt this form to our notation for this data,

$$CO_2(t) = a(b^t).$$

As an example, we derive the exponential equation whose graph passes through the points (50, 310.0) and (75, 331.0) which correspond to data in 1950 and 1975. In this example, we round the constant a to two places and the base b to four places. Substitution gives the two equations

$$310 = a(b^{50})$$
$$331 = a(b^{75}),$$

which we must solve for a and b. Divide the second equation by the first,

$$\frac{331}{310} = \frac{a(b^{75})}{a(b^{50})},$$

which eliminates a, and yields

$$b^{25} = 1.0677.$$

Solve this equation on your calculator to obtain

$$b = 1.0026.$$

Now substitute this value for b into the first equation and get

$$310 = a(1.0026^{50}),$$

and hence

$$a = 272.26.$$

The exponential equation is

$$CO_2(t) = 272.26(1.0026)^t,$$

where $CO_2(t)$ = atmospheric concentration in ppm in year $1900 + t$. Table 14.2 shows differences and error for this function.

TABLE 14.2

t	$CO_2(t)$	Actual CO_2 Concentration	Difference
0	272.3	280.0	7.7
50	310.0	310.0	0.0
75	330.8	331.0	0.2
80	335.1	338.5	3.4
90	343.9	354.0	10.1

Error: 21.4

Figure 14.2 shows the graph of this function with the original data points in their correct relative positions.

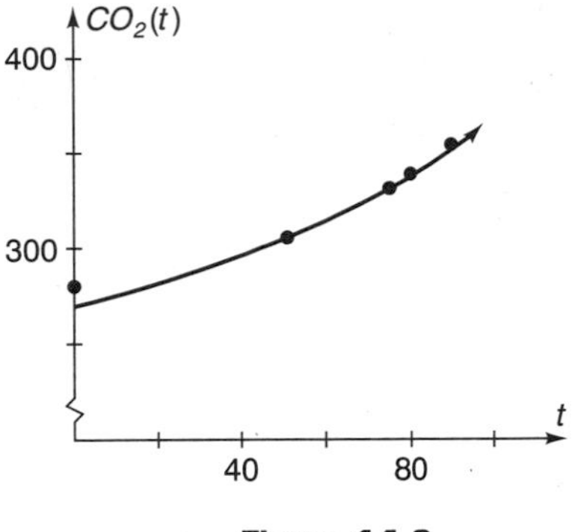

Figure 14.2

Carbon dioxide concentration is increasing as evidenced by the data, and of course this is reflected in the exponential equation

$$CO_2(t) = 272.26(1.0026)^t.$$

In fact in any one year period, the amount of increase is

$$CO_2(t + 1) - CO_2(t) = 272.26(1.0026^{t+1}) - 272.26(1.0026^t).$$

Factor out $272.26(1.0026^t)$ from this expression to get

$$272.26(1.0026)^t[1.0026 - 1] = 272.26(1.0026^t)(.0026).$$

This last expression can be used to compute the amount of increase in carbon dioxide concentration in any year $1900 + t$, and we leave it in this unsimplified form for a reason.

Next, we can see the percentage change by dividing the annual increase by the total concentration in year $1900 + t$ to get

$$\frac{272.26(1.0026^t)(.0026)}{272.26(1.0026^t)} = .0026.$$

This number, .0026, means that the CO_2 concentration increases by 0.26 percent each year. The number 0.26 percent is known as the *growth rate.* (*Growth rate* is the annual percentage change.) The growth rate can be easily determined in this manner for any exponential function. Note that for this kind of function, growth rate is independent of t.

Determine the remaining nine exponential equations using the CO_2 concentration data in Table 14.1. Determine the error for each and find the best function.

Use the best equation to determine each of the following:

1. CO_2 concentration in 2010;
2. the year when concentration will be 560 ppm, which is double the preindustrial level;
3. the growth rate. (Can you find a general formula for growth rate?)

In Chapter 3, you modeled U.S. population with a linear equation. Now, we look at world population, which requires a more sophisticated model. The table below shows population in billions worldwide in the indicated year. Model these data with an exponential equation using $t = 0$ in 1950; then use your model to answer the questions below.

Year	World Population ($\times 10^9$)
1950	2.6
1960	3.1
1970	3.7
1980	4.5
1990	5.3

Source: *Statistical Abstract of the United States, 1993*, Bureau of the Census.

1. Predict world population in the year 2020.
2. When will world population double its present size?
3. Determine the growth rate of world population.
4. Suppose that, as of this year, this growth rate is decreased by 10 percent. How would this effect the population in the year 2020? (In order to answer this question, write a new exponential equation using $t = 0$ in the present year and the new growth rate.)

EARTH NOTE Carrying Capacity

Biologists and other scientists are interested in how large the population of a species can be. In order to determine this, they compute the *biological carrying capacity* for a given species in a given locale. This *carrying capacity* is the maximum number of the species which its environment can support. Unchecked, population will increase to a number greater than its carrying capacity, then dramatically decrease due to the lack of sustenance from the environment. Often this decrease is so great that recovery is impossible.

Although the biological carrying capacity was developed for use for nonhuman species, some scientists in 1972 attempted to adapt this concept to global human population. The figure of six billion was reached, under the assumption that the population existed on a European living standard.

Use your worldwide population model to determine the year in which population will reach this figure.

Source: Kaufman, Donald, and Franz, Cecelia, *Biosphere 2000: Protecting our Global Environment,* HarperCollins College Publishers (New York, 1994).

14.2 CARBON EMISSION AND INCREASE IN CO_2 CONCENTRATION

When carbon is released into the air as carbon dioxide, some of it is reabsorbed by plants and the oceans as part of the carbon cycle, something like the way blood circulates. A portion of the carbon dioxide that is released into the air by human activities stays in the atmosphere. We can estimate the effect this has on climate if we can describe how carbon dioxide emitted from human sources accumulates in the atmosphere and how this affects the concentration of CO_2. We discuss this problem in terms of carbon instead of carbon dioxide since most available information is given this way. If we know the rate of growth of carbon emission (either from all human activity or from some particular source), it is not difficult to estimate how much that will raise the carbon dioxide concentration in the atmosphere. The study below predicts the worldwide carbon emissions by year beginning in 1991; then this model is used to determine how many gigatons stay in the atmosphere. Finally we convert gigatons to parts per million, the standard measure of carbon dioxide in the atmosphere.

It was estimated that the amount of carbon in 1991 in the atmosphere was 750 gigatons. A gigaton, abbreviated GT, is a billion metric tons (10^9 metric tons), and is a convenient unit to use. Also, in 1991 the annual emission was about 7 GT. In the last few years, the amount of carbon being emitted annually has been increasing at a rate of about 1 percent per year but during most of the eighties the amount of carbon being emitted grew at a rate more like 1.7 percent. A rate of 1 percent means that if 7 gigatons of carbon are emitted this year, then next year, the emissions will be 7.07 GT (the 1991 level plus an additional 1 percent of the 1991 level):

$$7 + 7(.01) = 7(1 + .01) = 7(1.01).$$

In the year following, another 1 percent increase yields

$$\begin{aligned} 7(1.01) + (.01)[7(1.01)] &= 7(1.01)(1 + .01) \\ &= 7(1.01)(1.01) \\ &= 7(1.01)^2 \end{aligned}$$

GT of carbon emitted. In general, if t is the number of years after 1991, the emission will be

$$7(1.01)^t \text{ GT carbon.}$$

Under the assumption that the 1991 carbon emission was 7 GT and that the growth rate continues to be 1 percent, we determine the increase in atmospheric carbon dioxide concentration due to worldwide carbon emission.

The equation which can be used to predict the amount of carbon emitted annually is

$$C(t) = 7(1.01)^t,$$

where $C(t)$ = gigatons of carbon emitted worldwide in year $1991 + t$. This function $C(t)$ describes annual emission.

Next, we use this carbon emission function to write a function which defines total emission for the period from 1991 through $1991 + t$. This function will be the sum

$$7 + 7(1.01) + 7(1.01)^2 + \cdots + 7(1.01)^t,$$

which is a geometric series with $a = 7$, $r = 1.01$, and $n = t$. Its sum is

$$\frac{7[(1.01)^{t+1} - 1]}{1.01 - 1} = \frac{7[(1.01)^{t+1} - 1]}{.01} = 700[(1.01)^{t+1} - 1].$$

This is the total amount of carbon emitted worldwide over this period. Although not all carbon emitted remains in the atmosphere, 58 percent of it does. (This estimate, .58, is known as the airborne fraction.) Hence the total carbon (in GT) emitted from 1991 through $1991 + t$ which remains in the atmosphere is

$$(.58)(700)[(1.01)^{t+1} - 1] = 406[(1.01)^{t+1} - 1].$$

Note that this expression only gives carbon accumulation from 1991 through $1991 + t$, but total atmospheric carbon can be obtained by adding to this quantity the estimated 1991 amount of carbon in the atmosphere, which is 750 gigatons. This yields the function

$$AC(t) = 750 + 406[(1.01)^{t+1} - 1].$$

where $AC(t)$ = total atmospheric carbon in gigatons in $1991 + t$.

This function can be used to predict the total atmospheric carbon in any year after 1991. For example, in year 2010,

$$t = 2010 - 1991 = 19,$$

so total atmospheric carbon will be

$$AC(19) = 839.4 \text{ gigatons.}$$

1. During most of the eighties, the amount of carbon emitted grew at a rate of 1.7 percent. Derive a function which defines total atmospheric carbon if this growth rate had continued. Assume other data remain the same.
2. Derive a similar function which defines total atmospheric carbon if the growth rate were reduced to .5 percent. Assume other data remain the same.

What we really want to know is what this accumulation of carbon means in terms of atmospheric concentration of CO_2. So we need to write an equation which defines *ppm* as a function of *AC*; this converts gigatons of atmospheric carbon into parts per million, the standard measure of atmospheric carbon dioxide concentration. We assume this equation to be linear, and specify two points which will determine this equation.

In 1991 the total carbon in the atmosphere was 750 GT and the concentration was 354 ppm. Certainly if there were no carbon, then the concentration would be 0 ppm. Now find the equation of the line through the points (0, 0) and (750, 354). The line has slope .47, and with variables (*AC*, *ppm*), the equation is

$$ppm(AC) = .47AC.$$

Now put the pieces together and determine the concentration as a function of time t. That is just the composite function:

$$\begin{aligned} ppm(AC(t)) &= .47[750 + 406(1.01^{t+1} - 1)] \\ &= 352.5 + 190.82(1.01^{t+1} - 1). \end{aligned}$$

Using this new model, we determine when atmospheric concentration will be twice the preindustrial level of 280 ppm. (Remember that this is when the temperature will increase about 4.5° F.) This can be done by solving the equation

$$560 = 352.5 + 190.82(1.01^{t+1} - 1).$$

Using the trace feature, we determine

$$t = 73 \text{ (rounded)}.$$

This predicts that the CO_2 concentration will double 73 years after 1991, or in the year 2064. Note that this is much earlier than our previous models predict; this could be due to recent increased carbon emission.

Use the $AC(t)$ functions with each of the growth rates which you derived in the previous exercise to determine the year in which this doubling of preindustrial CO_2 will occur under each new assumption.

EARTH NOTE Solar Electricity

Solar photovoltaic (PV) cells convert sunlight to electricity and are considered to be the cleanest source of energy. The cost of electricity from PV cells has dropped significantly over the last ten years.

The table below gives the amount of PV electricity generated (in megawatts) and the corresponding price, in 1986 dollars.

Year	Quantity (megawatts)	Price (1986 dollars)
1980	4.0	15
1982	8.0	10
1984	22.2	7
1986	24.6	4

1. Plot the quantity and corresponding price on a (q, p) coordinate system.
2. Find all exponential equations of the form $p = a(b^q)$ modeling this data. Compute the error for each and indicate the best equation. Graph the best equation on the same coordinate system as Exercise 1, and use this equation to answer the remaining questions.
3. What price would increase the demand to 26 megawatts?
4. What would the demand be for PV cells if the price drops to $3.50? to $3.00?

Source: Global Tomorrow Coalition. *The Global Ecology Handbook* (Boston: Beacon Press, 1990).

CHAPTER 15

Contribution to Atmospheric CO_2 Concentration from Three Sources

15.1 APPROXIMATING PREDICTED DATA WITH EXPONENTIAL FUNCTIONS

In this chapter we estimate the amount of carbon dioxide which will be added to the atmosphere from each of the three sources over the next ten years. One way to accomplish this is to evaluate each respective total emission function at t values corresponding to each year from 1995 through 2005, sum the answers, and multiply by the airborne fraction. This is a tedious method, particularly if a twenty- or thirty-year period were of interest. So, we will derive a new function which makes these computations quicker and easier.

Let $F(t)$ denote the total emission function for one of the three sources (if automobiles, $F(t) = CO_2A(t)$; if energy, $F(t) = TCE(t)$; if deforestation, $F(t) = CDF(t)$). The new function will be exponential, and will closely approximate $F(t)$. This you can see on your calculator when working through your respective section in this chapter. There are three advantages to using an exponential function:

1. the approximation to $F(t)$ is good;
2. a growth rate can easily be determined;
3. it is easy to sum the annual accumulations.

To derive the exponential model we need two points: compute $F(t)$ for values of t corresponding to the years 1995 and 2005. This gives the points which will be used to derive the exponential function $E(t)$ which approximates $F(t)$. You may think we are being inconsistent because we are only using two particular points to model emission. We are not finding all equations using all pairs of points and computing errors to find the best. There is a difference in this case; previously we have been modeling *real* data, and here we are approximating *predicted* data points.

We provide an example which illustrates this exponential approximation of data predicted from an emission function.

EXAMPLE 15.1

Approximate the parabolic function which predicts carbon emission from natural gas consumption over the time period 1995–2005 with an exponential function.

The natural gas emission function in gigatons carbon is

$$\begin{aligned} F(t) &= .01454NGC(t) \\ &= .01454(.0810t^2 - 2.5470t + 37.8100) \\ &= .0012t^2 - .0370t + .5498, \end{aligned}$$

where $t = 0$ in 1970, but $t \geq 10$ since this is the second part of a piecewise function. Use $F(t)$ to predict emission in the years 1995 and 2005:

$$\begin{aligned} F(25) &= .3748; \\ F(35) &= .7248. \end{aligned}$$

Now, derive the exponential equation $E(t) = a(b^t)$ which passes through the points (25, .3798) and (35, .7868). We get the two equations

$$\begin{aligned} .7248 &= a(b^{35}) \\ .3748 &= a(b^{25}) \end{aligned}$$

Divide the first equation by the second to eliminate a:

$$\begin{aligned} \frac{.7248}{.3748} &= \frac{a(b^{35})}{a(b^{25})} \\ 1.9338 &= b^{10} \\ b &= 1.0682. \end{aligned}$$

Substitute this value for b into the second equation to get

$$\begin{aligned} .3748 &= a(1.0682^{25}) \\ a &= .0720. \end{aligned}$$

Hence the exponential approximation is

$$E(t) = .0720(1.0682^t),$$

where $25 \leq t \leq 35$. Graph the functions $F(t)$ and $E(t)$ on the same coordinate system; set x min $= 20$, x max $= 40$, y min $= .3$, and y max $= .8$. This illustrates the close approximation of $F(t)$ by $E(t)$ over this time period.

In the group work which follows, it will be much easier to determine atmospheric accumulation of carbon (dioxide) if $t = 0$ in 1995 for the exponential emission functions. Therefore we continue with Example 15.1 and translate so that this occurs. Originally, $t = 0$ in 1970, $1995 - 1970 = 25$, so replace t in $E(t)$ by $t + 25$ to get the new function

$$\begin{aligned} E(t) &= .0720(1.0682^{t+25}) \\ &= .0720(1.0682^{t})(1.0682^{25}) \\ &= .0720(1.0682^{t})(5.2037) \\ &= .3747(1.0682^{t}), \end{aligned}$$

where $E_0(t)$ = predicted carbon emission from natural gas consumption in the year $1995 + t$, $0 \leq t \leq 10$. ▲

15.2 APPROXIMATING CONTRIBUTION FROM EACH SOURCE

Each group should now complete this study by following the steps listed on the next page. Each group should do the study for its chosen or assigned CO_2 source. Recall,

$$\begin{aligned} F(t) &= \text{original derived total emission function} \\ &\quad (F = CO_2A,\ F = TCE, \text{ or } F = CDF), \end{aligned}$$

and

$$E(t) = \text{exponential approximation to } F(t),$$

as described above.

PASSENGER CARS

1. Derive the exponential function $E(t)$ which approximates $CO_2A(t)$ over the years 1995–2005.
2. Simultaneously graph $CO_2A(t)$ and $E(t)$ on your calculator, with x-range set at x min corresponding to the t value for 1995, and x max corresponding to the t value for 2005. Here you should see the close approximation of $CO_2A(t)$ by $E(t)$.
3. Adjust the function $E(t)$ to get a new exponential function $E_0(t) = a_0b^t$, where t now equals 0 in 1995. This requires a substitution for t and simplification to obtain the form a_0b^t.
4. Determine the growth rate for CO_2 emission from automobiles. Recall, growth rate is annual percentage change, and hence is

$$\frac{a_0b^{t+1} - a_0b^t}{a_0b^t} = \frac{a_0b^t(b - 1)}{a_0b^t} = b - 1.$$

5. Compute the total CO_2 emission from automobiles during the time period 1995–2005. This is the sum

$$E_0(0) + E_0(1) + \cdots + E_0(10),$$

which is a geometric series (see Chapter 13).

6. Ultimately, you want to know the increase in atmospheric concentration due to automobiles. You will need to convert pounds of CO_2 to gigatons of carbon. Information necessary to do this is:

$$\text{weight } CO_2 = \text{weight carbon} \times 3.667;$$
$$1 \text{ lb} = .4545 \times 10^{-12} \text{ gigatons}$$

Determine the amount of carbon from automobiles that actually remains in the atmosphere in gigatons (use the airborne fraction). Then use the equation derived in Chapter 14, Section 14.2, to determine the resulting increase in atmospheric concentration (in ppm).

7. Suppose the growth rate of carbon emission which you computed in Step 4 on the previous page was reduced by 10 percent. This would give a new exponential function $E_1(t)$ which would be of the form $a_0 b_1^t$, where

$$b_1 - 1$$

is this new growth rate (note a_0 is the same as the a_0 in $E_0(t)$). Repeat Steps 5 and 6 to determine the increase in atmospheric carbon (in ppm) from your source with the reduced growth rate. Compare your two answers.

ENERGY CONSUMPTION

1. Derive the exponential function $E(t)$ which approximates $TCE(t)$ in the years 1995–2005.
2. Simultaneously graph $TCE(t)$ and $E(t)$ on your calculator, with x-range set at x min corresponding to the t value for 1995, and x max corresponding to the t value for 2005. Here you should see the close approximation of $ICE(t)$ by $E(t)$.
3. Adjust the function $E(t)$ to get a new exponential function $E_0(t) = a_0 b^t$, where t now equals 0 in 1995. This requires a substitution for t and simplification to obtain the form $a_0 b^t$.
4. Determine the growth rate for carbon emission from energy consumption. Recall, growth rate is annual percentage change, and hence is

$$\frac{a_0 b^{t+1} - a_0 b^t}{a_0 b^t} = \frac{a_0 b^t (b - 1)}{a_0 b^t} = b - 1.$$

5. Compute the total emission from energy consumption during the time period 1995–2005. This is the sum

$$E_0(0) + E_0(1) + \cdots + E_0(10),$$

which is a geometric series (see Chapter 13).

6. Ultimately, you want to know the increase in atmospheric concentration due to energy consumption: Determine the amount of carbon from this source which actually remains in the atmosphere in gigatons (use the airborne fraction). Then use the equation derived in Chapter 14, Section 14.2, to determine the resulting increase in atmospheric concentration (in ppm).

7. Suppose the growth rate of carbon emission which you computed in Step 4 on the previous page was reduced by 10 percent. This would give a new exponential function $E_1(t)$ which would be of the form $a_0 b_1^t$, where

$$b_1 - 1$$

is this new growth rate (note a_0 is the same as the a_0 in $E_0(t)$). Repeat Steps 5 and 6 to determine the increase in atmospheric carbon (in ppm) from your source with the reduced growth rate. Compare your two answers.

DEFORESTATION

1. Derive the exponential function $E(t)$ which approximates $CDF(t)$ for the years 1995–2005.

2. Simultaneously graph $CDF(t)$ and $E(t)$ on your calculator, with x-range set at x min corresponding to the t value for 1995, and x max corresponding to the t value for 2005. Here you should see the close approximation $CDF(t)$ by $E(t)$.

3. Adjust the function $E(t)$ to get a new exponential function $E_0(t) = a_0 b^t$, where t now equals 0 in 1995. This requires a substitution for t and simplification to obtain the form $a_0 b^t$.

4. Determine the growth rate for carbon emission from deforestation. Recall, growth rate is annual percentage change, and hence is

$$\frac{a_0 b^{t+1} - a_0 b^t}{a_0 b^t} = \frac{a_0 b^t (b - 1)}{a_0 b^t} = b - 1.$$

5. Compute the total emission from deforestation during the time period 1995–2005. This is the sum

$$E_0(0) + E_0(1) + \cdots + E_0(10),$$

which is a geometric series (see Chapter 13).

6. Ultimately you want to know the increase in atmospheric concentration due to deforestation. You will need to convert metrics to gigatons:

$$1 \text{ gigaton} = 10^9 \text{ metric tons.}$$

Determine the amount of carbon from this source which actually remains in the atmosphere in gigatons (use the airborne fraction). Then use the equation derived in Chapter 14, Section 14.2, to determine the resulting increase in atmospheric concentration (in ppm).

7. Suppose the growth rate of carbon emission which you computed in Step 4 on the previous page was reduced by 10 percent. This would give a new exponential function $E_1(t)$ which would be of the form $a_0 b_1^t$, where

$$b_1 - 1$$

is this new growth rate (note a_0 is the same as the a_0 in $E_0(t)$). Repeat Steps 5 and 6 to determine the increase in atmospheric carbon (in ppm) from your source with the reduced growth rate. Compare your two answers.

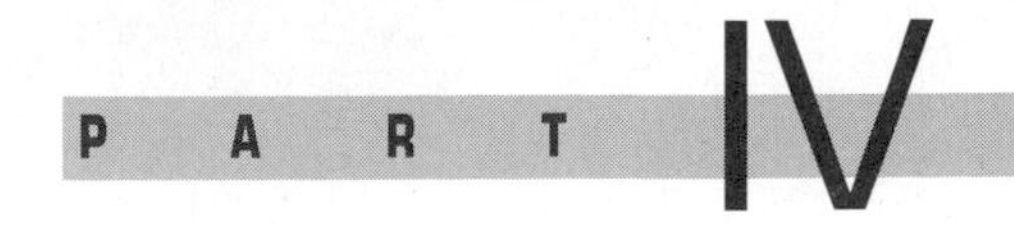

PART IV

Connecting Carbon Dioxide, People, and Money

Introduction

From our studies of cars, energy consumption, and deforestation, we have learned that CO_2 emission from each of these sources is sharply increasing; that is, the atmospheric concentration of carbon dioxide is becoming greater and greater. But, think for a minute: cars don't just drive around by themselves, healthy trees don't just fall down, and most animals don't need electric lights. People drive cars, people cut down rain forests, and people need electricity. In Chapter 16, we use models of U.S. and world population which were derived earlier to study connections between people and carbon dioxide.

The three sources of atmospheric carbon dioxide which we have studied also happen to be major factors in the U.S. economy, the health of which is sometimes measured by the gross domestic product, abbreviated GDP. This is the monetary value of all goods and services produced in the United States during a defined time period. Generally, it is a good measure of the state of the economy, but not always. So it would seem reasonable that there would be a relationship between GDP and things such as number of cars, some portion of deforestation, and energy consumption. In Chapter 17, we model GDP and write certain factors involved in our study of CO_2 emission as functions of GDP. In this way we can study the relationship between GDP and atmospheric CO_2. It also gives a relationship between money and global warming.

The mathematical topic necessary to conduct these studies is the concept of an inverse function. This was covered in Chapter 10.

A Reminder. Coefficients of functions and equations in all environmental chapters should be rounded as follows: linear, two places; quadratic, exponential, and logarithmic, four places. Sometimes we deviate from this practice but not without warning.

Some discrepancy may occur in answers if round-off is done during calculation as opposed to only rounding the final result. In all situations the final answer should be rounded to the same number of places as the original data.

CHAPTER 16

People

16.1 HOW MANY PEOPLE ARE THERE?

In this chapter, we study connections between population and carbon dioxide. First, we derive a relationship between U.S. population and atmospheric carbon dioxide concentration. (Even though U.S. population is only 4.7 percent of world population, it is responsible for 25 percent of CO_2 emissions from human activity.)

You derived the U.S. population model in Chapter 3, Section 3.1; that model is

$P(t) = 2.34t + 180.70$, where

$P(t) =$ total U.S. population in millions in $1960 + t$.

Also in Chapter 3, the atmospheric CO_2 concentration was derived as the linear model

$$CO_2C(t) = 1.44t + 280, \text{ where}$$
$$CO_2C(t) = \text{atmospheric } CO_2 \text{ concentration in ppm in year } 1939 + t.$$

We wish to study the relationship between U.S. population and CO_2 concentration, so a function which defines CO_2C in terms of U.S. population would be appropriate; that is, the function we need is $CO_2C(P)$. In order to determine this, we can find the inverse of the population function and form a composite with the CO_2 concentration function.

But first, there is a time discrepancy in the two equations; this needs to be adjusted so that both t's are 0 in the same year. In the population equation, $t = 0$ in 1960, whereas in the CO_2C equation, $t = 0$ in 1939 (a 21-year difference). We adjust so that $t = 0$ in 1939 for both. This involves replacing t by $t - 21$ in the $P(t)$ equation to get the new population model (which we still call $P(t)$)

$$P(t) = 2.34(t - 21) + 180.70$$
$$P(t) = 2.34t + 131.56$$

where now

$$P(t) = \text{U.S. population } (\times 10^6) \text{ in } 1939 + t.$$

Next, we find the inverse of this population equation. Replace $P(t)$ by P to get

$$P = 2.34t + 131.56$$

and solve for t:

$$2.34t = P - 131.56$$
$$t = \frac{P - 131.56}{2.34},$$

so

$$t = .43P - 56.22.$$

This is the inverse of the adjusted population equation, which we substitute for t in the CO_2C equation,

$$CO_2C(t) = 1.44t + 280$$
$$CO_2C(P) = 1.44(.43P - 56.22) + 280,$$

so

$$CO_2C(P) = .62P + 199.04.$$

The slope of this function means that an increase of one million people corresponds to an increase in CO_2 concentration of .62 ppm. (Remember that population P is in millions and CO_2C is in parts per million.)

We can also answer such interesting questions as: what population increase would correspond to an increase in an average global temperature of 1°F? In order to answer this, we recall our $GT(CO_2C)$ function from Chapter 5,

$$GT(CO_2C) = .016\ CO_2C - 4.48.$$

The slope of this line is .016 which means that a 1 ppm increase in CO_2 concentration corresponds to a .016°F increase in average global temperature, and so a .62 ppm increase in CO_2 concentration corresponds to a

$$(.016)(.62) = .01°\text{F}$$

increase in average global temperature. This means that an increase of one million people corresponds to an increase of .01°F in average global temperature. Hence the population increase which would correspond to a 1°F increase is

$$100(1 \text{ million}) = 100 \text{ million people.}$$

You can use the $P(t)$ function to predict how many years it will take for the present global temperature to increase 1°F.

Use the long-term exponential model for CO_2 from Section 14.1 and the exponential model for world population, also from Section 14.1, to express CO_2 concentration in terms of world population.

16.2 OF PEOPLE AND CO_2 SOURCES

Each group should read the section appropriate to its assigned or chosen source. In these studies, each group will relate population to one factor involved in CO_2 emission from its source.

Passenger Cars

As long as cars are readily available to us, the more people there are, probably the more cars there will be. So to see that relationship directly, you should study the connection between number of cars and population. To complete this study, follow the steps listed below.

1. Recall the original U.S. population model (Section 3.1)

$$P(t) = 2.34t + 180.70,$$

where $t = 0$ in 1960, and recall the $A(t)$ model for number of U.S. automobiles (Section 8.1). Adjust the above $P(t)$ function to get a new $P(t)$ function in which $t = 0$ in 1940 (base year for autos).

2. Find the inverse of the adjusted $P(t)$ function.
3. Form the composite function $A(P)$ which defines number of cars in terms of U.S. population.
4. Provide a verbal interpretation of the slope of $A(P)$.
5. Determine the present U.S. population, and the projected U.S. population ten years from now.
6. Determine CO_2 emission from automobiles this year and the projected emission ten years from now (use $CO_2A(t)$ from Chapter 12).
7. Suppose the rate of population growth slows by 10 percent, that is, from the current rate of 2.34×10^6 to 2.11×10^6 per year. What would be the number of automobiles corresponding to this decrease ten years from now? What effect would this have on CO_2 emission ten years from now?

Energy Consumption

As the population grows, more and more electricity will be used. A major portion of the electricity in the United States is generated from coal burning power plants, so it seems reasonable to study the relationship between the population and coal consumption. To do this study, follow the steps listed below.

GROUP WORK

1. Recall the original coal consumption function $CC(t)$ (Section 9.1). This is a piecewise function with the first part parabolic and the second part linear. The second part corresponds to the most recent years, and since we will be concentrating on recent and future time periods in this study, we only need consider that part of this piecewise function. Now, also recall the original U.S. population function (Section 3.1),

$$P(t) = 2.34t + 180.70,$$

where $t = 0$ in 1960. Adjust this function to get a new population function, also called $P(t)$, where $t = 0$ in 1970. This is necessary so that the t in both $CC(t)$ and $P(t)$ coincide.
2. Find the inverse of the adjusted $P(t)$ function.
3. Form the composite function $CC(P)$, which defines coal consumption in the U.S. in terms of U.S. population.
4. Write a verbal interpretation of the slope of $CC(P)$.
5. Determine the present U.S. population and the projected U.S. population ten years from now.
6. Determine carbon emission from energy consumption this year and the projected emission ten years from now (use $TCE(t)$ from Chapter 12).
7. Suppose the rate of population growth slows by 10 percent, that is, from the current rate of 2.34×10^6 to 2.11×10^6 per year. What would be the corresponding coal consumption with this decrease in ten years? What effect would this have on carbon emission from energy consumption in the U.S. ten years from now?

Deforestation

As the population increases, so does the demand for goods, in particular for furniture, homes, and many other wood products. The United States is an importer of wood cut from rain forest, so a study of the relationship between population and deforestation for logging seems reasonable. To do this study, follow the steps listed below.

1. Recall the original U.S. population function (Section 3.1)

$$P(t) = 2.34t + 180.70,$$

where $t = 0$ in 1960, and recall the logging function $L(t)$ from Section 11.1. Adjust the above function $P(t)$ to get a new population function $P(t)$ in which $t = 0$ in 1940 (base year for deforestation). This is necessary so that the t in both $P(t)$ and $L(t)$ coincide.

2. Find the inverse of the adjusted $P(t)$ function.
3. Form the composite function $L(P)$ which defines hectares of rain forest cut for logging in terms of population of the United States.
4. The function $L(P)$ is, of course, not linear so its slope is not defined. You can, however, use the slope concept to approximate the rate of increase of logging with respect to population. Here's how: determine the U.S. population P_0 this year and evaluate $L(P)$ for this number, that is, $L(P_0)$. Then evaluate $L(P)$ for this year's population plus one million, that is, $L(P_0 + 1)$. The difference $L(P_0 + 1) - L(P_0)$ is a "slope approximation." (Recall that if $y = f(x)$ is a linear function, then if x increases by 1 unit, the change in y is the slope. Also recall that the population P is given in millions, so increasing the population by one million means $P + 1$.) Now, write a verbal interpretation of the current rate of increase of logging in the rain forest with respect to population.

5. Determine the present U.S. population and the projected population ten years from now.
6. Determine carbon emission due to deforestation this year and the projected emission ten years from now (use $CDF(t)$ from Chapter 12).
7. Suppose the rate of growth of the U.S. population slows by 10 percent; that is, from the current rate of 2.34×10^6 to 2.11×10^6 per year. What would be the number of hectares cut for logging corresponding to this decrease ten years from now? What effect would this have on carbon emission due to deforestation ten years from now?

EARTH NOTE American Indian Population

Did you know that the American Indian population declined from an estimated 5 million in 1492 to approximately 237,000 in 1900? However, the Native Americans have made a strong comeback: as of the 1990 census, they have increased their number to almost 2 million.

The table below provides the American Indian population (AIP) in thousands in the indicated year. The 1492 figure is an estimate by historian Russell Thornton, and the remaining data are obtained from the U.S. Census Bureau.

Year	$AIP \times 10^3$
1492	5000
1890	248
1930	343
1980	1364
1990	1878

1. Make points out of these data; use $t = 0$ in 1890 (so $t = -398$ in 1492), the year of the tragic Wounded Knee massacre.
2. Model these data with a piecewise function: linear for 1492–1890, and parabolic for years after 1890.
3. Graph your model on a (t, AIP) coordinate system, and plot the original data points in their correct position relative to the function.

   Use your model to answer Exercises 4–8.
4. In what year was Indian population at a minimum, and what was that population?
5. In 1876, Custer and his army were defeated by a mixed band of tribes at Little Big Horn, and the Sioux lost their Black Hills in the forced Treaty of 1876. Use your model to estimate the Indian population at that time.

Columbus "discovered" America in 1492 marking the beginning of the influx of nonIndians to the North American continent. Over the period 1492–1890 the nonIndian population grew from practically nil to 62,700,000 in the geographical U.S. alone.

6. Assuming a linear growth, write the equation which approximates this growth of nonIndians. Use $t = 0$ in 1890.
7. Determine the year in which American Indian population would have been equal to the nonIndian population.
8. Determine the inverse of the nonIndian population function over the period 1492–1890, and then form a composite to find the function which defines American Indian population as a function of nonIndian population from 1492 to 1890.
9. Write a verbal interpretation to the slope of the linear function determined in Exercise 8 above.

Money

17.1 MODELING GROSS DOMESTIC PRODUCT

The gross domestic product (GDP) is the total output of goods and services produced by the domestic economy. In this section, we will model GDP and write certain factors involved in our study of CO_2 emission as functions of GDP. This way we can study the relationship between GDP and CO_2 in the atmosphere. It also gives a relationship between money and global warming.

Table 17.1 gives GDP in constant 1987 dollars by year. (Constant 1987 dollars means that the figures are adjusted as if the dollar had constant purchasing power.) Figure 17.1 is a (t, GDP) coordinate system with corresponding points plotted.

TABLE 17.1 Gross Domestic Product

Year	(Constant 1987 Dollars × 10^{12})
1960	1.971
1970	2.874
1980	3.776
1990	4.878
1992	4.923

Source: *Statistical Abstract of the United States, 1993*, Bureau of the Census.

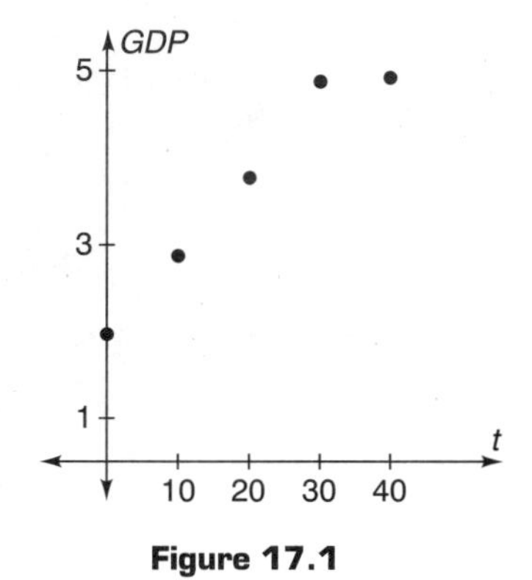

Figure 17.1

EARTH NOTE Consumer Waste

Did you know that on the average, an American consumer spends $225 each year on the packaging of the goods purchased? On top of this, we throw away 180 million plastic razors each year, and discarded aluminum cans could be used to build 6000 DC-10 airliners.

Source: Kaufman, Donald, and Franz, Cecelia, *Biosphere 2000: Protecting our Global Environment,* Harpercollins College Publishers (New York, 1994).

1. Model the data in Table 17.1 using an exponential equation; name your best function $GDP(t)$, where $t = 0$ in 1960.
2. Recall the total CO_2 concentration model from Chapter 14,

$$CO_2(t) = 280(1.0024^t),$$

where $t = 0$ in 1900. Adjust the function $CO_2(t)$ to obtain a new $CO_2(t)$ function where $t = 0$ in 1960.
3. Find the inverse of the $GDP(t)$ function from Step 1.
4. Form the composite function $CO_2(GDP)$, which defines CO_2 as a function of GDP.
5. Graph the function $CO_2(GDP)$.
6. The function $CO_2(GDP)$ is not linear, so its slope is not defined. However, you can use the concept of the slope to approximate the rate of increase of CO_2 with respect to GDP. (Some of you may have already done something like this in Section 16.2.) Here's how: determine the gross national product GDP_0 this year and evaluate $CO_2(GDP)$ for this number; that is, $CO_2(GDP_0)$. Then evaluate $CO_2(GDP)$ for this year's gross national product plus 1 billion dollars, that is, $CO_2(GDP_0 + 1)$. The difference $CO_2(GDP_0 + 1) - CO_2(GDP_0)$ is a "slope approximation." (Recall that for a linear function $y = f(x)$, if x increases one unit then the slope tells you how much y changes; also recall that since GDP is expressed in billions of dollars, increasing GDP by one billion means $GDP + 1$). Now write a verbal interpretation of the current rate of increase of CO_2 with respect to GDP.

17.2 GDP AND CO_2 CONTRIBUTORS

Car Groups

Do you think that the GDP affects the gasoline consumed in the United States? Probably, since we drive around to malls to spend those dollars, drive to work, and to the beach if it's still there. In this study we examine the relationship between GDP and gasoline consumption from automobiles.

Recall the function $M(t)$ from Section 8.3; to complete this study, follow the steps listed below.

1. Adjust the $GDP(t)$ function to get a new $GDP(t)$ function where $t = 0$ in 1940.
2. Find the inverse of the adjusted $GDP(t)$ function.
3. Write the composite $M(GDP)$, which defines average number of miles driven in terms of GDP (use the second part of the piecewise function $M(t)$).
4. Graph the function $M(GDP)$.
5. Use your $GDP(t)$ function to determine the current GDP. Use this to compute the "slope approximation" for $M(GDP)$ this year, and write a verbal interpretation.

GDP may also be related to the number of cars on the road. Recall the function $A(t)$ from Section 8.1.

6. Write the composite $A(GDP)$, which defines number of automobiles in the United States in terms of GDP.
7. Compute the "slope approximation" for $A(GDP)$ for the current year, and write a verbal interpretation.

Energy Groups

GDP and energy consumption are probably related. A lot of energy is required to keep up with the increasing demand for production of goods. In this study we examine the relationship between GDP and energy consumption. Recall the total energy consumption function

$TEC(t)$, $t = 0$ in 1970 (Section 9.4). To complete this study, follow the steps below.

1. Adjust the $GDP(t)$ function to get a new $GDP(t)$ function where $t = 0$ in 1970.
2. Find the inverse of the adjusted $GDP(t)$ function.
3. Write the composite $TEC(GDP)$, which defines total United States energy consumption in terms of GDP.
4. Graph the function $TEC(GDP)$.
5. Use your $GDP(t)$ function to determine the current GDP. Use this to compute the "slope approximation" for $TEC(GDP)$ for this year, and write a verbal interpretation.

Rain Forest Groups

Even though most of the rain forests are not in the United States, there is a relationship between what we consume here and the acreage cut in Brazil or Thailand. Your hamburger may come from cattle which graze on cut rain forest land, and your teak table may come from rain forest in Thailand. In this study we examine the relationship between GDP and carbon emission from deforestation. Recall the functions $L(t)$ and $CG(t)$, where $t = 0$ in 1940 (Sections 11.1 and 11.2). To complete this study, follow the steps listed below.

1. Adjust the $GDP(t)$ function to get a new $GDP(t)$ function where $t = 0$ in 1940.
2. Find the inverse of the adjusted $GDP(t)$ function.
3. Write the composite $L(GDP)$, which defines the amount of rain forest cut for logging in terms of GDP.
4. Graph the function $L(GDP)$.

5. Use your $GDP(t)$ function to determine the current GDP. Use this to compute the "slope approximation" for $L(GDP)$ for this year, and write a verbal interpretation.
6. Write the composite $CG(GDP)$, which defines the amount of rain forest cut for cattle grazing in terms of GDP.
7. Graph the function $CG(GDP)$.
8. Compute the "slope approximation" for $CG(GDP)$ for this year, and write a verbal interpretation.

EARTH NOTE National Income Distribution

Did you know that in the United States, only 20 percent of the population receives almost 50 percent of the national income?

Source: Worldwatch Institute, *State of the World, 1993,* W. W. Norton & Company (New York, London, 1993).

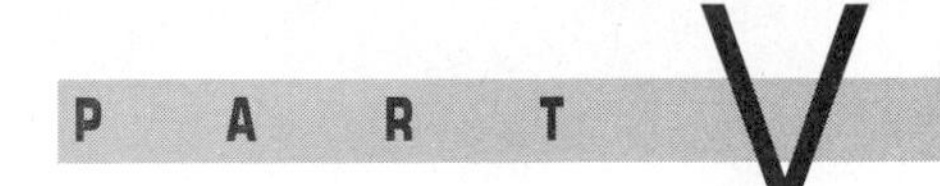

Alternate Energy and New Trends: Reducing Carbon Dioxide Emission

Introduction

You have, at this point, completed a fairly thorough study of carbon dioxide emission from three major sources: automobiles in the U.S., power consumption, and deforestation in the tropics. You have also seen the effect of the overall accumulation of CO_2, in particular, the predicted rise in ocean levels and the possible resulting land loss.

In the following chapters, we will study ways to decrease the emission of carbon dioxide in order to slow the resulting temperature increase. Chapter 19 studies the use of alternative energy sources and other ways of using traditional ones in order to reduce CO_2 emission, and Chapter 20 invites students of *Earth Algebra* to formulate their own solutions to the problem of carbon dioxide emission.

New mathematical topics needed for Part V are linear inequalities and linear programming, which are covered in Chapter 18. Any of the mathematical concepts and methods studied in this text can and should be incorporated into your own plan for CO_2 reduction.

A Reminder. Coefficients of functions and equations in all environmental chapters should be rounded as follows: linear, two places; quadratic, exponential, and logarithmic, four places. Sometimes we deviate from this practice but not without warning.

Some discrepancy may occur in answers if round-off is done during calculation as opposed to only rounding the final result. In all situations the final answer should be rounded to the same number of places as the original data.

Linear Inequalities in Two Variables and Systems of Inequalities

18.1 GRAPHING LINEAR INEQUALITIES IN TWO VARIABLES

A linear inequality in the two variables x and y looks like

$$ax + by \leq c,$$

or

$$ax + by < c,$$

or

$$ax + by \geq c,$$

Note: Chapter 18 is a prerequisite for Chapters 19 and 20.

or

$$ax + by > c,$$

where a, b, and c are constants. A solution to an inequality is any pair of numbers x and y which satisfy the inequality.

The rules for finding the solution set of a linear inequality are much the same as those for finding the solution to a linear equation:

1. Add or subtract the same expression to both sides;
2. Multiply or divide both sides by the same nonzero quantity; if that quantity is negative, then the inequality must be reversed.

Here's an example.

EXAMPLE 18.1

Determine the solution set of $5x + 2y \leq 17$.

One solution to this is $x = 2$ and $y = 3$ because

$$5(2) + 2(3) = 16,$$

which is indeed less than or equal to 17. A pair of numbers which does not form a solution is $x = 3$ and $y = 2$, because

$$5(3) + 2(2) = 19,$$

which is not less than or equal to 17. The pair $x = 2$ and $y = 3$ isn't the only solution; as a matter of fact, there are infinitely many solutions.

Since we can't write down all possible solutions to a linear inequality, a good way to describe the set of solutions to any linear inequality is by a graph. If the pair of numbers x and y is a solution, then think of this pair as a point in the plane, so the set of all solutions can be thought of as a region in the xy-plane. We continue Example 18.1 in order to illustrate how to determine this region. First, solve the inequality for y in terms of x.

$$5x + 2y \leq 17$$

$$2y \leq -5x + 17$$

$$y \leq -\frac{5x}{2} + \frac{17}{2}.$$

Next, graph the line

$$y = -\frac{5x}{2} + \frac{17}{2}.$$

The set of points (x, y) which lie on this line is the set of all (x, y) such that y is exactly equal to $-\frac{5x}{2} + \frac{17}{2}$. These points make up part of the set of solutions to the inequality, but not all. We see that y can also be less than $-\frac{5x}{2} + \frac{17}{2}$, so all points below the line would also be solutions. The shaded region in Figure 18.1 shows the solution set. ▲

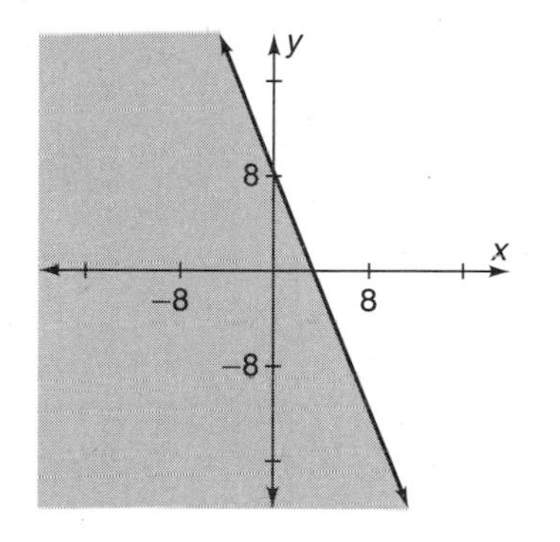

Figure 18.1

EXAMPLE 18.2

Graph the solution set for the inequality

$$3x - 8y \geq 12.$$

Solve for y:

$$-8y \geq -3x + 12$$

$$y \leq \frac{3x}{8} - \frac{3}{2}.$$

Do you notice that the inequality is reversed? That's because we divided by a negative number—any time you multiply or divide an inequality by a negative number, you must reverse the inequality.

Next, graph the line

$$y = \frac{3}{8}x - \frac{3}{2}.$$

Points on this line are part of the solution set, the other part consists of all points below the line. See the shaded region in Figure 18.2. ▲

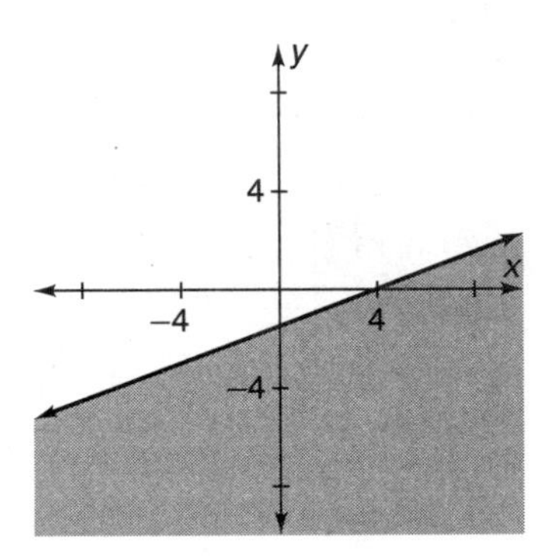

Figure 18.2

EXAMPLE 18.3

Graph the solution set for the inequality

$$-10x - 2y > 7.$$

Solvng for y gives

$$-2y > 10x + 7$$

$$y < -5x - \frac{7}{2}. \quad \text{(Remember the minus?)}$$

This example is a little different because there's no "equals mark" in the inequality. But you still graph the line

$$y = -5x - \frac{7}{2},$$

except draw it as a dashed line. This indicates that the line itself is not part of the solution set. The actual solution set consists of all points below the dashed line. This is because y must be strictly less than $-5x - \frac{7}{2}$. See Figure 18.3. ▲

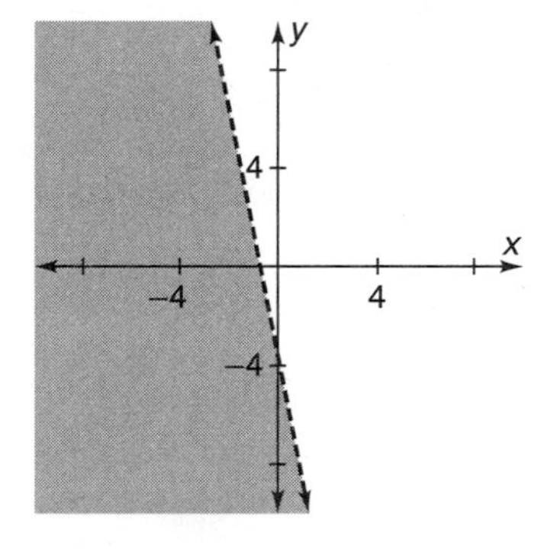

Figure 18.3

18.1 THINGS TO DO

For Exercises 1–5, graph the solution set to each inequality.

1. $5x + 2y \geq 12$

2. $2x - 8y < 6$

3. $0.1x - 2.5y \geq 1.4$

4. $12x - 11y \leq 65$

5. $10.2x + 4.5y \leq 14.1$

18.2 SYSTEMS OF LINEAR INEQUALITIES IN TWO VARIABLES

A system of linear inequalities is a set of one or more linear inequalities; a solution is a pair of numbers (x, y) which satisfies all of the inequalities.

EXAMPLE 18.4

Determine the solution set to the system of linear inequalities

$$x + 5y \leq 20$$
$$3x + 2y \leq 21.$$

The pair of numbers $x = 1, y = 2$ is one solution because

$$1 + 5(2) = 11 \leq 20$$
$$3(1) + 2(2) = 7 \leq 21.$$

The pair $x = 0, y = 5$ is not a solution because it doesn't even satisfy the first inequality

$$0 + 5(5) = 25,$$

which is not less than or equal to 20. Notice that it does satisfy the second inequality, but in order to be a solution, it must satisfy both.

As before, a system can have an infinite number of solutions, so we present its solution set by a region in the plane. To illustrate this, we continue with Example 18.4,

$$x + 5y \leq 20$$
$$3x + 2y \leq 21.$$

Solve each inequality for y:

$$y \leq -\frac{1}{5}x + 4$$

$$y \leq -\frac{3}{2}x + \frac{21}{2}.$$

Graph each of the lines on the same coordinate system. The solution set for the first inequality lies in the region below and on the line $y = -\frac{1}{5}x + 4$ and the solution set for the second inequality lies in the region on and below the line $y = -\frac{3}{2}x + \frac{21}{2}$. The solution set for the system lies in the region common to both, and is the darker region shown in Figure 18.4.

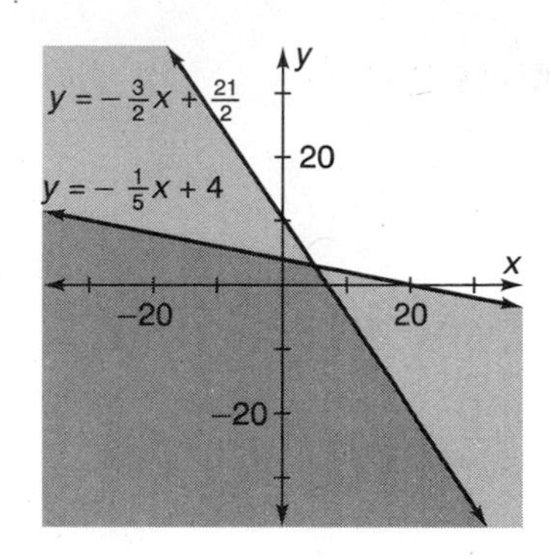

Figure 18.4

The "corner" of this region is the intersection of the two lines, and "corners" will be very important in the next chapter. In order to find this point, return to the original system of inequalities and replace the "inequalities" with "equal marks." The resulting two equations are the equations of the two lines

$$x + 5y = 20$$
$$3x + 2y = 21.$$

This is a system of two linear equations in two unknowns, and its solution is precisely the point of intersection of the two lines. You know how to solve this system by setting up the coefficient matrix

$$A = \begin{bmatrix} 1 & 5 \\ 3 & 2 \end{bmatrix},$$

and constant matrix

$$B = \begin{bmatrix} 20 \\ 21 \end{bmatrix},$$

and solving on your calculator. The solution matrix is

$$A^{-1}B = \begin{bmatrix} 5 \\ 3 \end{bmatrix},$$

so $x = 5$ and $y = 3$. The point of intersection, or corner of the region, is $(5, 3)$. ▲

EXAMPLE 18.5

Graph the solution set to the system

$$7x - 5y \leq 12$$

$$2x + 3y \leq 18$$

$$x \geq 0$$

$$y \geq 0$$

and find all corners.

The last two inequalities in this system simply restrict the solution set to points in the first quadrant (that's the quadrant in which both coordinates are nonnegative). Solve the first two inequalities for y to get

$$y \geq \frac{7}{5}x - \frac{12}{5}$$

$$y \leq -\frac{2x}{3} + 6.$$

The solution set to the first inequality lies in the region on and above the graph of the line $y = \frac{7}{5}x - \frac{12}{5}$, and the solution set to the second lies on and below the line $y = -\frac{2x}{3} + 6$. The common region which is in the first quadrant is the solution set to the original four inequalities, and is shown in Figure 18.5.

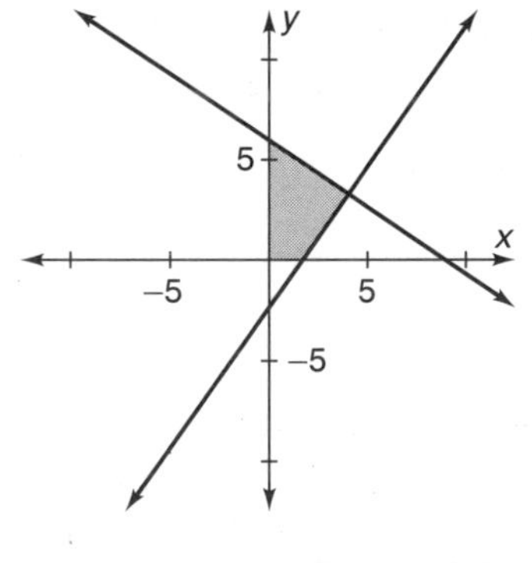

Figure 18.5

Next we find corners of this region. There are four of them. One of them is really easy; that's the origin: $x = 0$ and $y = 0$. Two others are also easy; these are the ones on the axes. The one on the x-axis is the x-intercept of the line $7x - 5y = 12$. To find this, take $y = 0$, and solve for x, $x = \frac{12}{7} = 1.71$, so this corner is $(1.71, 0)$. On the y-axis, the corner is the y-intercept of the line $2x + 3y = 18$; take $x = 0$ to get $y = 6$, so the corner is $(0, 6)$. The last point requires some work, but not much. You must find the intersection of the lines $7x - 5y = 12$ and $2x + 3y = 18$, which is the solution to the linear system

$$7x - 5y = 12$$
$$2x + 3y = 18.$$

The coefficient matrix is

$$A = \begin{bmatrix} 7 & -5 \\ 2 & 3 \end{bmatrix}$$

and the constant matrix is

$$B = \begin{bmatrix} 12 \\ 18 \end{bmatrix}.$$

The solution is

$$A^{-1}B = \begin{bmatrix} 4.06 \\ 3.29 \end{bmatrix},$$

so $x = 4.06$ and $y = 3.29$, and the corner is (4.06, 3.29). See Figure 18.6. ▲

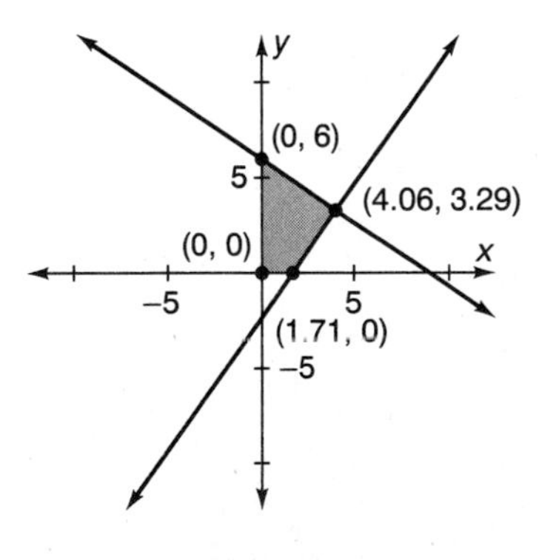

Figure 18.6

EXAMPLE 18.6

Graph the solution set to the system

$$
\begin{aligned}
x + 4y &\le 20 \\
3x + 4y &\ge 28 \\
x - y &\le 7
\end{aligned}
$$

and find all corners.

As before, graph each line and shade the appropriate region; the solution set to the entire system is the region common to all three, and is shown in Figure 18.7, complete with corners. ▲

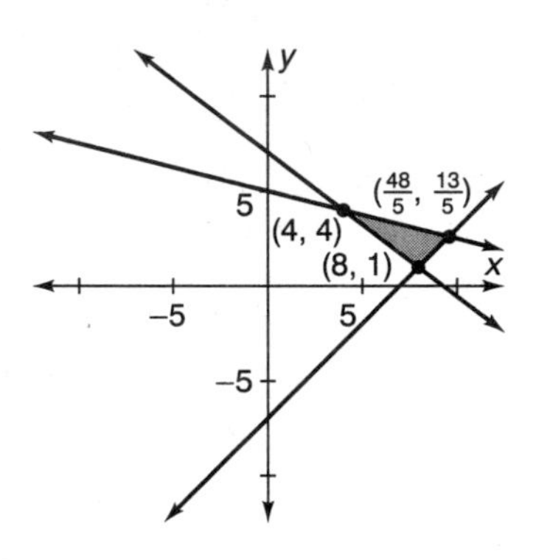

Figure 18.7

18.2 THINGS TO DO

For Exercises 1–5, graph the solution set to each system of linear inequalities, and find all corners.

1. $x + y \ge 10$
$2x - y \ge 2$

2. $100x + 400y \le 750$
$75x + 250y \le 700$
$x \ge 0$
$y \ge 0$

3. $x - y \le 10$
$2x + y \ge 12$

4. $1.5x + 2.4y \le 3.9$
$3.2x + 10.2y \le 14.1$
$x \ge 0$
$y \ge 0$

5. $x - y \ge 0$
$x - 2y \le 2$
$x + y \le 2$

18.3 LINEAR PROGRAMMING

The primary reason for studying systems of linear inequalities at this point is to determine solutions to mathematical problems which can be solved using *linear programming* methods. Specifically, one has a function which is to be maximized or minimized, but the solution is subject to certain restrictions. The function itself is called the *objective function,* and the restrictions are called *constraints* and usually appear in a system of inequalities. In order to solve a linear programming problem, follow the steps below.

1. Graph the inequalities which express the constraints and determine the solutions set as a region in the plane; this region is called the *feasible region.* In *Earth Algebra,* all feasible regions will be bounded, which insures the existence of maximum and minimum values for the objective function.
2. Determine the corners of the feasible region.

3. The maximum or minimum of the objective function will occur at one of these corners. In order to find which one, evaluate the objective function at each corner; the maximum is the largest value and the minimum is the smallest.

Here are three examples.

EXAMPLE 18.7

Maximize the objective function

$$A = 3x + 5y$$

subject to the constraints

$$2x + y \leq 16$$
$$x + 3y \leq 18$$
$$x \geq 0$$
$$y \geq 0.$$

First, graph each line and shade the appropriate region; then find corners. This is shown in Figure 18.8.

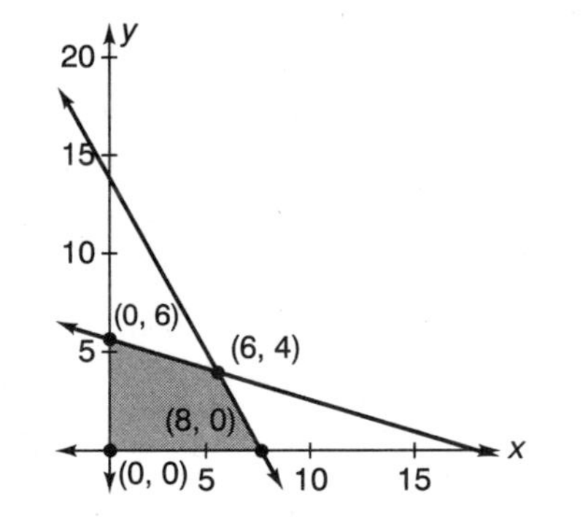

Figure 18.8

Now evaluate the objective function at each corner

Corner	Functional Value
(0, 0)	$A = 3(0) + 5(0) = 0$
(0, 6)	$A = 3(0) + 5(6) = 30$
(6, 4)	$A = 3(6) + 5(4) = 38$
(8, 0)	$A = 3(8) + 5(0) = 24$

Thus, the maximum is 38 and occurs at corner (6, 4). ▲

Here is a note on why you only need to look at the corners when you are finding the maximum or minimum value of the objective function. Of course, $3x + 5y$ can be as large as you want, but the problem is to stay inside the feasible region. Figure 18.9 shows $3x + 5y = 50$, $3x + 5y = 40$, and the one that passes through the corner at which the solution occurs. The latter line is $3x + 5y = 38$. None of the points on the line $3x + 5y = 50$ are inside the feasible region, so none of these points are valid solutions. Notice that the line $3x + 5y = 40$ is closer and all the lines $3x + 5y =$ any number are parallel. Slide the highest line down staying parallel until you first hit a point in the feasible region. What point did you hit? A corner! So where is the maximum? At a corner! (See Figure 18.9.)

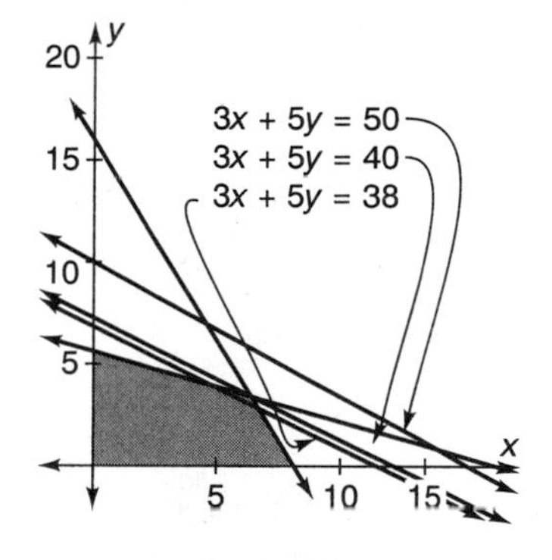

Figure 18.9

EXAMPLE 18.8

Maximize the objective function

$$B = 8x + y$$

subject to the same constraints in Example 18.7.

The feasible region with corners are shown in Figure 18.8. Evaluate this objective function at the corners.

Corner	Functional Value
(0, 0)	$B = 8(0) + 0 = 0$
(0, 6)	$B = 8(0) + 6 = 6$
(6, 4)	$B = 8(6) + 4 = 52$
(8, 0)	$B = 8(8) + 0 = 64$

The maximum for the objective function B is 64, and occurs at corner (8, 0). ▲

EXAMPLE 18.9

Minimize the objective function

$$C = 11x + 10y$$

subject to the constraints

$$x + y \geq 15$$
$$6x + 5y \geq 80$$
$$4x + 3y \leq 60$$
$$x \geq 0$$
$$y \geq 0.$$

The graph of the feasible region with its corners is shown in Figure 18.10.

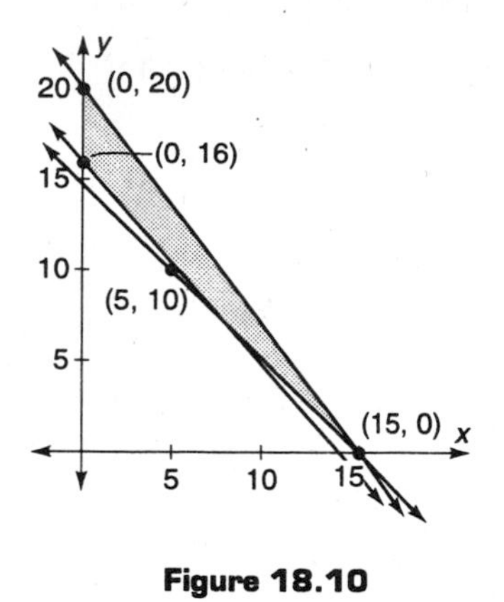

Figure 18.10

Evaluate the objective function at the corners.

Corner	Functional Value
(5, 10)	$C = 11(5) + 10(10) = 155$
(0, 16)	$C = 11(0) + 10(16) = 16$
(0, 20)	$C = 11(0) + 10(20) = 200$
(15, 0)	$C = 11(15) + 10(0) = 165$

The minimum value of the objective function is 155 which occurs at (5, 10). ▲

18.3 THINGS TO DO

1. Maximize the objective function

$$E = x + y$$

subject to the constraints

$$2x + 3y \leq 6$$
$$x \geq 0$$
$$y \geq 0.$$

2. Maximize the function

$$F = x - 2y$$

subject to
$$\begin{aligned} y - x &\le 4 \\ 4x + y &\le 24 \\ x &\ge 0 \\ y &\ge 0. \end{aligned}$$

3. Minimize the function

$$G = 3x + 2y$$

subject to
$$\begin{aligned} x + y &\le 8 \\ y &\ge 2x - 4 \\ 4x + y &\ge 8. \end{aligned}$$

4. Maximize the function

$$H = 4x + y$$

subject to
$$\begin{aligned} 2y - x &\le 2 \\ y + 3x &\le 15 \\ x + y &\ge 1 \\ y &\ge 0. \end{aligned}$$

5. Minimize the function

$$J = 2x - y$$

subject to
$$\begin{aligned} x + y &\ge 5 \\ x &\ge y - 5 \\ x &\le 7 \\ y &\ge 0. \end{aligned}$$

Cost and Efficiency of Alternate Energy Sources

The techniques used in the studies in this chapter are known to mathematicians as *linear programming* methods. (This subject is treated in Chapter 18 but some repetition is included here.) Generally, a linear programming problem involves a particular function, whose maximum (or minimum) value is of interest. But other restrictions apply to the situation; these appear as inequalities. The function to be maximized (or minimized) is called the *objective function,* and the inequalities are called the *constraints.* The studies below seek to maximize utilization subject to cost restriction and imposed carbon dioxide emission levels.

In general there are four steps involved in setting up and solving a linear programming problem.

Step 1. Determine the objective function.

Step 2. Determine the constraints. (In the study below, this involves using previously developed models to decide reasonable limitations.)

Step 3. Determine and graph the feasible region with corners.

Step 4. Determine the maximum or minimum of the objective function.

19.1 METHANOL VERSUS GASOLINE

In this section, we will look at how gasoline burning cars and methanol burning cars affect the environment. Methanol emits less CO_2, costs less per gallon, but is less fuel efficient (that is, you get fewer miles per gallon). Suppose some cars burn gasoline and others burn methanol. We will show you how to answer the following question: what combination will let the total population drive the most number of miles without exceeding a fixed cost or emitting more than a fixed amount of carbon dioxide?

First you need some information:

1. when burned, gasoline emits 20 lbs. of CO_2 per gallon;
2. when burned, methanol emits 9.5 lbs. of CO_2 per gallon;
3. it takes 75 percent more methanol than gasoline to go one mile;
4. the average cost of gasoline is $1.00 per gallon (Source: Buddy's Gas Station, N. Highland, Atlanta, GA); and
5. the average cost of methanol is $.60 per gallon.

(Numbers 4 and 5 are assumptions based on information current at the time of this writing.)

Note: the fuel efficiency difference is what makes this a nontrivial problem. Otherwise the solution is easy—switch to methanol! A gallon of methanol costs less and emits less CO_2, but it takes more to go each mile.

Step 1. **Objective Function.** We are trying to maximize the total number of miles that can be driven in the United States in a particular year, so we need the function that describes this. There are two choices of fuel which we consider, gasoline and methanol, and the overall problem is to determine the number of miles that should be driven on each in order to obtain this maximum (subject to cost and emission limitations). The variables in this study are

x = number of miles driven on gasoline (in billions),

y = number of miles driven on methanol (in billions).

Hence the function to be maximized is

$$TMD = x + y,$$

where *TMD* is the total number of miles ($\times\ 10^9$) that will be driven by all cars in the U.S. in a particular year. This is the *objective function.*

Step 2. **Constraints.**

A. Limitations. We first discuss the limitations of cost and emission. What limitations would be reasonable? In order to decide this, we look slightly into the future (somewhat arbitrarily, we choose 1998) to see what cost and emission are predicted by our derived models (Chapter 8) which are based on gasoline usage only.

The total number of miles driven by all cars in 1998 ($t = 58$) is projected to be:

$$M(58) \times A(58) \times 10^9 = 2168 \times 10^9;$$

the fuel efficiency is predicted to be:

$$MPG(58) = 24.5;$$

and therefore the number of gallons of gasoline consumed is

$$\frac{2168 \times 10^9}{24.5} = 88 \times 10^9.$$

At a cost of *$1 per gallon,* total fuel expenditures are $\$88 \times 10^9$. With an emission of 20 lbs. of CO_2 per gallon, total emission is 1760×10^9 lbs. of CO_2.

Our goal is to emit a little less CO_2, say 1600×10^9 lbs., at not much more expenditure, say $\$95 \times 10^9$. This gives two constraints. One constraint is a CO_2 constraint:

$$\text{total } CO_2 \leq 1600 \times 10^9 \text{ lbs.};$$

and the other is a cost constraint:

$$\text{total cost} \leq 95 \times 10^9 \text{ dollars.}$$

B. Constraints. Recall your answers to Exercises 5, 7, 9, and 10 at the end of Section 8.2.

5. Gallons of gasoline per mile = 0.0408
7. Pounds of CO_2 per mile (gasoline) = .8160
9. Gallons of methanol per mile = 0.0714
10. Pounds of CO_2 per mile (methanol) = 0.6783

If x miles are driven on gasoline, and y miles are driven on methanol, then total CO_2 emission from both is

$$0.8160x + 0.6783y \text{ (in billions of pounds).}$$

Our imposed limitation on CO_2 emission is 1600×10^9 lbs, so the constraint is expressed by the inequality

$$0.8160x + 0.6783y \leq 1600.$$

The second constraint is cost. At a cost of $1.00 per gallon, gasoline costs $0.0408 per mile. Methanol, at a cost of $.60 per gallon, will cost

(.60)(0.0714) = \$0.0428 per mile. So if x miles are driven on gasoline, and y miles are driven on methanol, then the total cost will be

$$.0408x + .0428y,$$

and with the imposed cost limitation of 95×10^9 dollars, this constraint is expressed by the inequality

$$.0408x + .0428y \leq 95.$$

There are two more constraints: $x \geq 0$ and $y \geq 0$. (You wouldn't drive a negative number of miles, now would you?) The table below (19.1) summarizes our work.

TABLE 19.1

	CO_2 per Mile (lbs)	Cost per Mile	(\$) Miles ($\times 10^9$)
Gasoline	.8160	.0408	x
Methanol	.6783	.0428	y
Limitations		1600($\times 10^9$)	95($\times 10^9$)

Here is the problem:

Maximize the function

$$TMD = x + y$$

subject to the constraints

$$.8160x + .6783y \leq 1600,$$

$$.0408x + .0428y \leq 95,$$

$$x \geq 0, y \geq 0.$$

That means that we have to find, among all the values for x and y that satisfy the inequalities, those which make $TMD = x + y$ as large as possible.

Step 3. **Feasible Region.** Up until now, all our work has been just setting up the problem; finally, we are ready to find its solution. First, graph the solution set to the system of inequalities (constraints) and find the corners (see Figure 19.1). The coordinates of the corners are rounded to the nearest integer.)

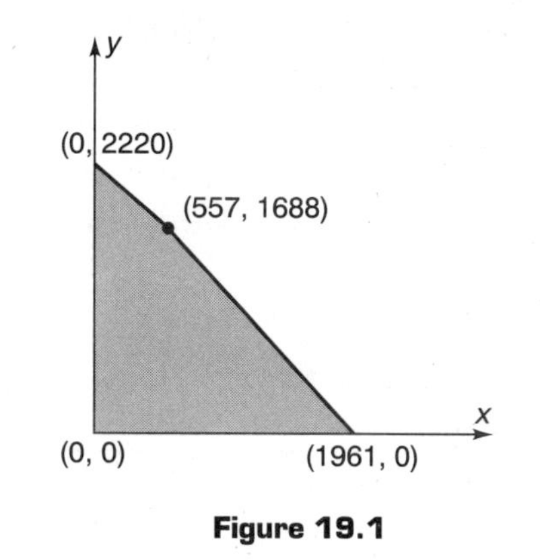

Figure 19.1

The shaded area, called the *feasible region,* shows which points satisfy all the constraints. That's a lot of points, and you have to figure which one makes $x + y$ biggest.

Step 4. **Maximum of the Objective Function.** The maximum value of the objective function, if there is a maximum, occurs at one of the corners.

Which corner point gives the greatest value for the objective function

$$TMD = x + y?$$

Substitute the values for x and y at each corner to get these answers:

$$0 + 2220 = 2220$$
$$557 + 1688 = 2245$$
$$1691 + 0 = 1691$$
$$0 + 0 = 0$$

The maximum value occurs at the corner (557, 1688). What does that mean? If methanol and gasoline are both available for use in U.S. automobiles, and if 557×10^9 miles are traveled on gasoline and 1688×10^9 miles on methanol, U.S. drivers can get the maximum number of total miles, which is 2245×10^9 miles. This can be done without exceeding the limit of 1600×10^9 lbs. of CO_2 emission and 95×10^9 dollars for fuel.

GROUP WORK

1. Now here's one for you that's almost the same. Using the same information as above, maximize the miles which can be driven if total costs are allowed to rise to $\$100 \times 10^9$ and CO_2 emissions can't exceed 1650×10^9 pounds.
2. Now try this: you want to improve the rate of CO_2 emission by 10 percent of the projected level for 1998, so you set a limit of 1590×10^9 lbs. of CO_2. To do this, you are willing to pay a little more for fuel, and you set a limit of $\$100 \times 10^9$. To further encourage the use of alternative fuels, you (having great power) raise the price of gasoline to $1.10 per gallon while keeping the price of methanol at $.60 per gallon. What is the maximum number of miles which could be driven?

GROUP WORK

There is a trade-off between cost and CO_2 emission when electric cars are compared to conventional cars. The information below (from the Knight-Ridder/Tribune News, printed in the *Atlanta Constitution,* Sept. 1991) gives a basis for comparison of gasoline powered versus electric vans.

On the average, a van is driven 8000 miles per year. The cost per mile to operate a full-sized gasoline powered van is $.368, while the per mile cost of an electric van is $.392. The gasoline powered van emits .690 kg of carbon dioxide per mile, while the electric powered van emits only .315 kg per mile. For the following problems, assume that you are currently operating a fleet of one hundred gasoline powered vans, and your fleet average miles driven is 8000 miles per year.

1. What is the current driving cost? current CO_2 emission?
2. Suppose that in order to decrease the CO_2 emissions by at least 10 percent, you decide that you will have a fleet which consists of both gasoline powered and electric powered vans, and to achieve this savings in CO_2 emission, you are willing to increase your costs by 10 percent. What is the new acceptable cost and new target CO_2 emissions?
3. What is the maximum size for the new fleet that will meet these conditions?

EARTH NOTE Power Needs for the United States

It is estimated that all United States power needs could be met by solar panels spread over 59,000 square kilometers, or equivalently by fully using the potential from wind power in three states: North Dakota, South Dakota, and Texas. Both wind and solar power are considered to be intermittent sources, but the peak demand periods in northern states in the winter coincide with the highest winds in the great plains, while the peak demands for air conditioning occur during the sunniest times of year.

Source: Worldwatch Institute, *State of the World,* 1994, W. W. Norton & Company (New York, London, 1994).

19.2 WHAT ARE SOME OTHER ALTERNATIVES?

Here is another study which can be resolved using methods of linear programming. This is to be a group effort.

You can use the techniques of linear programming to determine how to reduce the amount of carbon emitted from the consumption of power by changing from coal to natural gas. Almost half of the energy consumed in the United States is from these two sources. Annual coal consumption in 1995 is estimated to be 21.00×10^9 million BTU, while natural gas consumption is estimated to be 24.76×10^9 million BTU. Coal emits more carbon than natural gas, but it's cheaper. For every million BTU of coal consumed, 25 kilograms of carbon are emitted, while the equivalent consumption of natural gas emits only 14.5 kilograms of carbon. If there were no economic considerations, and an unlimited supply of natural gas, it would make sense to switch entirely from coal to natural gas. But there are economic considerations, and one is the relative prices. A million BTU of coal costs $1.413, while a million BTU of natural gas costs $1.538. Here's what we're going to do: first, we'll see what the current situation is, with regard to power consumption, carbon emission, and cost.

In order to determine reasonable limitations to impose on costs and carbon emission, we look at the situation in 1995, then determine how much power we can use if we restrict carbon emission but are willing to pay a little more. This can be determined by complcting Table 19.2A.

TABLE 19.2A 1995 DATA

Fuel	Total Consumption (10^9 million BTU)	Cost per Million BTU	Total Cost ($\$ \times 10^9$)	Carbon Emission per Million BTU	Total Carbon (kg $\times 10^9$)
Coal	21.00	$1.413	?	25 kg	?
Natural Gas	24.76	$1.538	?	14.5 kg	?
		Total Cost:	?	Total Emission:	?

After completing Table 19.2A you see that total cost is $\$66.7539 \times 10^9$ and total carbon emission is $884.02 \text{ kg} \times 10^9$.

Let's decrease the carbon emission by approximately 10 percent to 800×10^9 kg, and accept an increase in cost of about 10 percent to $\$75 \times 10^9$. We're ready to state the problem. Table 19.2B should help you.

TABLE 19.2B

Fuel	Cost ($ per Million BTU)	Carbon (kg per Million BTU)	Consumption (Million BTU $\times 10^9$)
Coal	1.413	25.0	x
Natural Gas	1.538	14.5	y
Limitations	75($\times 10^9$)	800($\times 10^9$)	

What is the maximum amount of energy from coal and natural gas that can be consumed subject to the constraints that the cost is no more than $\$75 \times 10^9$ and the carbon emitted is at most 800×10^9 kg? Assume that you can't change either the prices or the amount of carbon emitted per million BTU. Do this just like we did the methanol/gasoline problem. Round coordinates of corners to the nearest tenth.

What would happen if the price of coal goes up to $1.50 per million BTU and the total cost is allowed to reach $\$80 \times 10^9$? Round coordinates of corners to the nearest tenth.

CHAPTER 20

Creating New Models for the Future

Two students from the University of Texas at Austin built this solar car for a contest.

20.1 A NEW MODEL FOR FUTURE CARBON DIOXIDE EMISSION

In this chapter, you have the opportunity to formulate your own plan to decrease projected carbon dioxide emission. We show you one example of how this might work, and then suggest other similar ways to accomplish this. The following example was done in 1994 and reductions in projected CO_2 emission were made for the following ten year period.

What if you had the power to slow the increase of atmospheric carbon dioxide concentration so that the actual emission is 10 percent less in 2004 than the predicted amount for that year? That's a ten year period from 1994, so there is time to accomplish this relatively smoothly. You can't just command everyone to stop driving, or maybe just drive halfway to work, or to stop using electric lights immediately; that obviously will not work. So however you slow emission, it will have to be a gradual process over this ten year period. One simple way to accomplish this goal is to initiate a simple linear reduction of CO_2 emission from 1994 to the year 2004. Here's what we mean. Pretend you are very, very powerful and can somehow control CO_2 emission from energy consumption in the U.S., but you are not powerful enough to do anything about rain forest destruction, or anything else that produces carbon dioxide. So, what you can do is gradually, and uniformly, change CO_2 emission from energy consumption in the U.S. so that by the year 2004 a 10 percent reduction of the predicted overall emission will have been achieved. This may still be an increase over the 1994 emission, but at least it will be less than the predicted amount for 2004.

You need two things to figure out how this reduction will work. The first is the equation for worldwide carbon emission (Chapter 14, Section 14.2, translated so that $t = 0$ in 1994).

$$C(t) = 7.2121(1.01)^t,$$

where $C(t)$ = gigatons of carbon emitted in year 1994 + t, $t \geq 0$. The other is an equation which can be used to predict carbon emission in the U.S. from energy consumption (sources are coal, petroleum, natural gas, etc.):

$$EC(t) = 0.0030t^2 + .1010t + 2.5280,$$

where $EC(t)$ = gigatons of carbon emitted in 1994 + t, $t \geq 0$. (This model has been derived by the authors, and is provided for free.)

Next, to make up the difference between $C(t)$ and $EC(t)$, introduce a new variable, $OC(t)$, defined to be gigatons of carbon emitted from all sources other than energy consumption in the U.S., where t is the number of years after 1994, but $t \geq 0$.

This is nice because we get this simple equation

$$* \quad EC(t) + OC(t) = C(t).$$

You want to reduce $C(10)$ by 10 percent by changing the $EC(t)$ term in the equation.

The predicted carbon emission in 2004 is $C(10) = 7.9666$ gigatons. Reducing this figure by 10 percent is the same as multiplying it by .90 (why?), so we get

$$.90(7.9666) = 7.1699 \text{ gigatons of carbon.}$$

This is your target worldwide emission for 2004. You need to know $OC(10)$ also:

$$\begin{aligned} OC(10) &= C(10) - EC(10) \\ &= 7.9666 - 3.8380 \\ &= 4.1286 \text{ gigatons.} \end{aligned}$$

This figure must remain constant; you cannot change this.

In equation * we need to replace the $EC(t)$ term by the unknown target emission from energy consumption in the United States in 2004. Denote this unknown by y. To find this target emission y you must solve the equation

$$y + 4.1286 = 7.1699$$

to get

$$y = 3.0413 \text{ gigatons.}$$

It is interesting to note that this is a reduction of approximately 21 percent in the predicted emissions from energy for 2004.

A straight linear reduction from present EC, that is, $EC(0)$, means this: first $EC(0) = 2.5280$, which corresponds to the point $(0, 2.5280)$.

Next, $y = 3.0413$, which corresponds to $(10, 3.0413)$. Write the equation of the line which goes through these two points.

$$\text{Slope } m = \frac{3.0413 - 2.5280}{10 - 0} = 0.0513.$$

Substitute into the point-slope equation for a line:

$$TE - 2.5280 = 0.0513(t - 0)$$
$$TE(t) = 0.0513t + 2.5280,$$

where $TE(t) =$ target carbon emission from energy consumption in gigatons for year $1994 + t, 0 \leq t \leq 10$. (Here we keep four decimal places.)

Note that the slope of this linear function is .0513, which is positive, so carbon emission from energy is still growing, but at a smaller rate than previously predicted. You may say that .0513 is a really small number, and so emission is not increasing much. But, remember, we're talking about gigatons of carbon, and a gigaton is BIG!

This last restriction, $0 \leq t \leq 10$, is made because your new equation $TE(t)$, is only applicable during the ten years 1994–2004.

You can now use this reduction equation to project target emission for each year during the relevant time period. Here are the figures for alternate years (Table 20.1).

TABLE 20.1 Target Carbon Emission from Energy Consumption in U.S.

Year	Gigatons of Carbon
1995	2.5793
1997	2.6819
1999	2.7845
2001	2.8871
2003	2.9897

Figure 20.1 shows the graph of the original emission function together with the new reduction function.

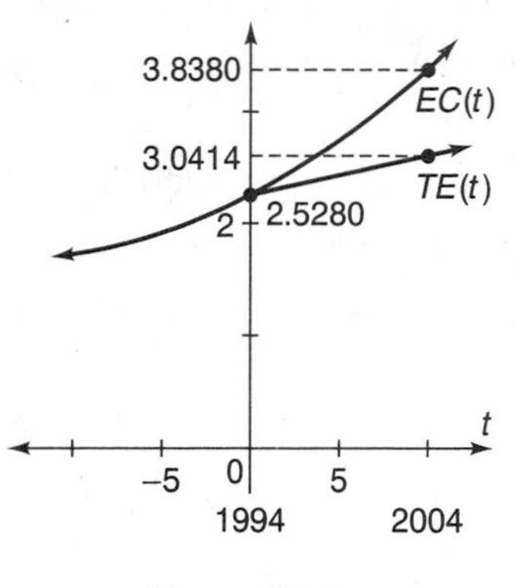

Figure 20.1

20.2 BE PART OF THE SOLUTION, NOT THE PROBLEM

Each group has been assigned (or chosen) one of the sources: automobiles, power, or deforestation. You also have your models which describe total CO_2 or carbon emission from each. Each model involves three factors: "automobiles" has number of cars, miles driven, and MPG; "power" has coal, natural gas, and petroleum; "deforestation" has logging, cattle, and agriculture.

Your group should present a ten year plan to reduce by 21 percent the predicted carbon or carbon dioxide emission from your source. Be sure to begin your reduction in the current year. One way to do this is with a straight linear change of one of the factors involved as shown in Section 20.1 so that in the ten years, CO_2 or carbon from your source is 21 percent less than its predicted value. You can change only one factor, the others will remain the same. For example, if your source is "automobiles," then you may choose to place requirements on the factor MPG; that is, require that MPG be increased uniformly over the next ten years so as to achieve the 21 percent reduction in emission in automobiles. You only change your MPG equation; the "number of cars" and "miles driven" equations will not change.

There are, of course, other ways to reduce carbon emission. Your group may decide to involve more than one factor or to use other types of equations. Be creative.

We suggest that your group prepare both a written and an oral presentation of your ten year plan. Both should include, at least, information such as that provided in the example we presented in Section 20.1.

The following items should be considered in devising your plan and preparing your report: feasibility, originality, clarity of explanation, and mathematical content.

There are many other interesting things you can do with your new reduction model. Be creative, think of your own.

Appendices

Appendix A
A Guide to the Variables of *Earth Algebra*

Appendix B
A Short Table of Conversions

Appendix C
The Quadratic Formula and Complex Numbers

Appendix D
The Sum of a Finite Geometric Series

Appendix E
Answers to Odd-Numbered Things to Do

A Guide to the Variables of *Earth Algebra*

For quick reference, each of the variables used to define a function in *Earth Algebra* is listed here with its definition. These include only the variables specifically used in the studies of environmental topics, not those which appear in the "prerequisite" chapters. The variables are listed in the order in which they appear in the text, and the page number refers to the page on which the variable is first introduced.

CO_2C = atmospheric CO_2 concentration (in ppm) by year, page 40 (linear model)

GT = average global temperature increase (in degrees Fahrenheit), page 58

OL = average ocean level increase (in feet), page 60

A = number ($\times 10^6$) of automobiles in the United States by year, page 114

MPG = average fuel efficiency (miles per gallon) per automobile in the United States by year, page 116

GPM = gallons of gasoline burned per mile by year, page 124

M = average number of miles ($\times 10^3$) each automobile in the United States drives per year, page 127

CC = coal consumption (in quads) in the United States by year, page 132

PC = petroleum consumption (in quads) in the United States by year, page 134

NGC = natural gas consumption (in quads) in the United States by year, page 135

TEC = total energy consumption (in quads) in the United States by year, page 137

L = number of hectares ($\times 10^6$) of rain forest cut for logging by year, page 166

CG = number of hectares ($\times 10^6$) of rain forest cut for cattle grazing by year, page 169

AD = number of hectares ($\times 10^6$) of rain forest cut for agriculture and development by year, page 173

CO_2A = total CO_2 emission (in lb. $\times 10^9$) due to automobiles in the United States by year, page 176

TCE = total carbon emission (in gigatons) due to energy (from coal, natural gas, and petroleum) consumption in the United States by year, page 178

CDF = total carbon emission (in metric tons) due to deforestation by year, page 179

CO_2 = atmospheric CO_2 concentration (in ppm) by year, page 190 (exponential model)

C = worldwide carbon emission (in gigatons) by year, page 197

AC = total atmospheric carbon accumulation (in gigatons) by year, page 197

ppm = parts per million of atmospheric carbon dioxide, used to convert gigatons of atmospheric carbon to ppm, page 198

P = United States population ($\times 10^6$) by year, page 213

GDP = United States gross national product (in dollars $\times 10^{12}$) by year, page 222

EC = total carbon emission (in gigatons) due to energy consumption in the United States by year, page 258

OC = total carbon emission (in gigatons) due to all sources other than energy consumption in the United States, page 258

TE = target carbon emission (in gigatons) from energy consumption in the United States in future years, page 260

A Short Table of Conversions

1 hectare = 2.47 acres

1 ton = 2000 lbs.

1 gigaton = 10^9 metric tons

1 metric ton = 1000 kilograms

1 lb. = .454 kilograms

1 quad = 10^{15} BTU

weight CO_2 = weight carbon × 3.667

1 gallon of gasoline burned emits 20 lbs. CO_2

1 quad of coal burned emits .02500 gigatons of carbon

1 quad of natural gas emits .01454 gigatons of carbon

1 quad of petroleum emits .02045 gigatons of carbon

1 hectare of destroyed rain forest emits approximately $\frac{1}{2}$ ton of carbon

The Quadratic Formula and Complex Numbers

The quadratic formula provides all solutions to quadratic equations. Its derivation is relatively simple using the method of completing the square. Consider the standard form for a quadratic equation,

$$ax^2 + bx + c = 0.$$

First, divide by the coefficient of x^2 to get

$$x^2 + \frac{b}{a}x + \frac{c}{a} = 0;$$

transfer the constant $\frac{c}{a}$ to the right to get

$$x^2 + \frac{b}{a}x = -\frac{c}{a}.$$

Now, completing the square on the left side of this equation means adding an appropriate term so that the result factors as a perfect square, that is, has the form $(x - h)^2$. This appropriate term is obtained by squaring half of $\frac{b}{a}$; that is, add $(\frac{b}{2a})^2$ to the left, and of course, to the right side too. This gives

$$x^2 + \frac{b}{a} + \frac{b^2}{4a^2} = \frac{b^2}{4a^2} - \frac{c}{a}.$$

Simplify the right and factor the left to get

$$\left(x + \frac{b}{2a}\right)^2 = \frac{b^2 - 4ac}{4a^2}.$$

Remember, we are solving the equation for x, so now take the square root of both sides:

$$x + \frac{b}{2a} = \pm\frac{\sqrt{b^2 - 4ac}}{2a};$$

finally solve for x and simplify to get the celebrated *quadratic formula,*

$$x = \frac{-b \pm \sqrt{b^2 - 4ac}}{2a}.$$

The quantity $b^2 - 4ac$, which is under the radical, is significant. If $b^2 - 4ac > 0$, there are two real solutions; if $b^2 - 4ac = 0$, there is only one real solution; if $b^2 - 4ac < 0$, there is no real solution.

The method of completing the square can be used to solve any quadratic equation should you forget the quadratic formula, or if it seems easier.

We discuss the special case where $b^2 - 4ac < 0$ in more detail. As pointed out earlier, there will be no real solution to the quadratic equation. However, there is a type of number, called a complex number (sometimes known as imaginary), which will be a solution. Define

$$\sqrt{-1} = i,$$

and a *complex number* to be one which can be written in the form $\alpha + \beta i$, where α and β are real numbers.

The next two examples illustrate how complex numbers are solutions to quadratic equations.

EXAMPLE 1

Solve $x^2 + 1 = 0$.

This is easier without using the quadratic formula. Simply transfer the 1 to the right to get

$$x^2 = -1$$
$$x = \pm\sqrt{-1},$$

so

$$x = \pm i.$$

Thus the equation $x^2 + 1 = 0$ has no real solution, but has two complex solutions. ▲

EXAMPLE 2

Solve $x^2 + 2x + 2 = 0$.

The quadratic formula gives

$$x = \frac{-2 \pm \sqrt{-4}}{2}.$$

To simplify, write $\sqrt{-4} = \sqrt{4}\sqrt{-1} = 2i$, so

$$x = \frac{-2 \pm 2i}{2} = -1 \pm i. \quad ▲$$

We complete this appendix with a brief discussion of operations on complex numbers. They can be added, subtracted, multiplied, and divided according to the following rules.

1. Addition: $(\alpha + \beta i) + (\gamma + \delta i) = (\alpha + \gamma) + (\beta + \delta)i$.
2. Subtraction: $(\alpha + \beta i) - (\gamma + \delta i) = (\alpha - \gamma) + (\beta - \delta)i$.
3. Multiplication: $(\alpha + \beta i)(\gamma + \delta i) = \alpha\gamma + \beta\gamma i + \alpha\delta i + \beta\delta i^2$
$= (\alpha\gamma - \beta\delta) + (\beta\gamma + \alpha\delta)i,$

or use "FOIL" and simplify.

4. Division. This requires a little work. The *conjugate* of the complex number $\alpha + \beta i$ is $\alpha - \beta i$. One significance of the conjugate is that the product of $\alpha + \beta i$ and its conjugate is a nonzero real number (provided $\alpha + \beta i \neq 0$):

$$(\alpha + \beta i)(\alpha - \beta i) = \alpha^2 + \beta^2.$$

Back to division of complex numbers:

$$\frac{\gamma + \delta i}{\alpha + \beta i} = \frac{(\gamma + \delta i)(\alpha - \beta i)}{(\alpha + \beta i)(\alpha - \beta i)} = \frac{(\alpha\gamma + \beta\delta) + (\alpha\delta - \beta\gamma)i}{\alpha^2 + \beta^2}$$

$$= \frac{\alpha\gamma + \beta\delta}{\alpha^2 + \beta^2} + \frac{\alpha\delta - \beta\gamma}{\alpha^2 + \beta^2} i.$$

Thus division is accomplished by multiplying numerator and denominator by the conjugate of the denominator, thus converting the denominator to a real number.

Here is an example of each of these operations:

1. Addition: $(3 - 2i) + (-1 + 5i) = 2 + 3i$;
2. Subtraction: $(7 - 6i) - (3 - i) = 4 - 5i$;
3. Multiplication: $(2 + 3i)(-1 - 4i) = -2 - 3i - 8i - 12i^2$

$$= -2 - 11i + 12$$

$$= 10 - 11i;$$

4. Division: $\dfrac{1 - 2i}{3 + 4i} = \dfrac{(1 - 2i)(3 - 4i)}{(3 + 4i)(3 - 4i)} = \dfrac{-5 - 10i}{9 + 16}$

$$= \frac{-5}{25} - \frac{10}{25} i = -\frac{1}{5} - \frac{2}{5} i.$$

This completes the real numbers.

APPENDIX D

The Sum of a Finite Geometric Series

Given the geometric series

$$a + ar + ar^2 + \cdots + ar^n.$$

with sum S, we have

$$S = a + ar + ar^2 + \cdots + ar^n.$$

Multiply both sides of this equation by r to get

$$rS = ar + ar^2 + \cdots + ar^n + ar^{n+1};$$

then arrange both equations as shown below and subtract the first from the second. The result is shown below the line.

$$\begin{array}{l} S = a + ar + ar^2 + \cdots + ar^n \\ \underline{rS = ar + ar^2 + \cdots + ar^n + ar^{n+1}} \\ rS - S = ar^{n+1} - a \end{array}$$

Now factor each side of the equal sign to get

$$S(r - 1) = a(r^{n+1} - 1),$$

and finally divide sides by $r - 1 (r \neq 1)$ to get

$$S = \frac{a(r^{n+1} - 1)}{r - 1}$$

which is the formula for the sum of the given geometric series. (Note, if $r = 1$, then $S = (n + 1)a$.

Answers to Odd-Numbered Things to Do

CHAPTER 1

Section 1.3 (page 15)

1. $f(2) = 1;\quad f(-3) = 11;$

$f(1.01) = 2.98;\quad f\left(-\frac{1}{2}\right) = 6$

3. $h(2) - 7;\ h(-3) = -1.1579;$

$h(1.01) = .6553;$

$h\left(-\left(-\frac{1}{2}\right)\right) = .4722$

5. $T(2.4) = -119.988;$

$T(24) = -119.88;$

$T(240) = -118.8;$

$T(2400) = -108;\quad T(24{,}000) = 0$

7. $H(2) = -1;$

$H(-3.3) = -1.5722;$

$H(11.001) = 32.5455;$

$H(5.5)$ undefined

9. $R(0) = 7.9373;\quad R(-2) = 8.1976;$

$R(30) = 0;\quad R(-1.52) = 8.1358;$

$R(50) =$ no real solution

CHAPTER 2

Section 2.1 (page 21)

1. $y = -1.4x + 0;$ slope $= -1.4;$

y-intercept: $(0, 0)$

3. $y = -2x + .9;$ slope $= -2;$

y-intercept: $(0, -9)$

5. $y = -\frac{3}{5}x + \frac{18}{5};$

slope $= -\frac{3}{5};$

y-intercept: $\left(0, \frac{18}{5}\right)$

7. $y = \frac{8}{5}x - 2;$ slope $= \frac{8}{5};$

y-intercept: $(0, -2)$

9. $y = -3$; slope $= 0$;
y-intercept: $(0, -3)$

Section 2.2 (page 25)

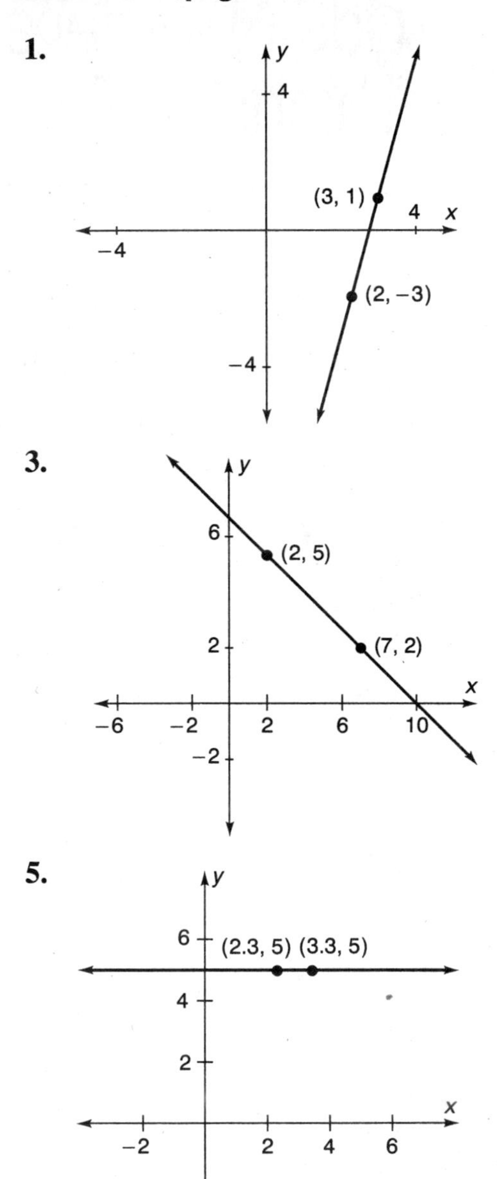

7.

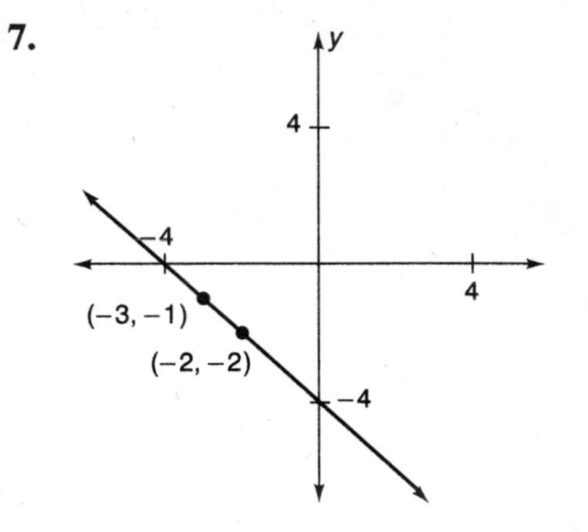

Section 2.3 (page 28)

1. x-intercept: $(6, 0)$;
y-intercept: $(0, 4)$; slope: $-\frac{2}{3}$

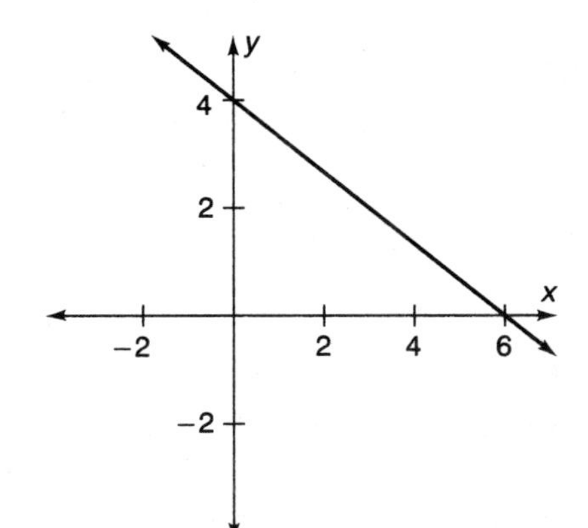

3. x-intercept: (0, 0);
y-intercept: (0, 0); slope: 1.5

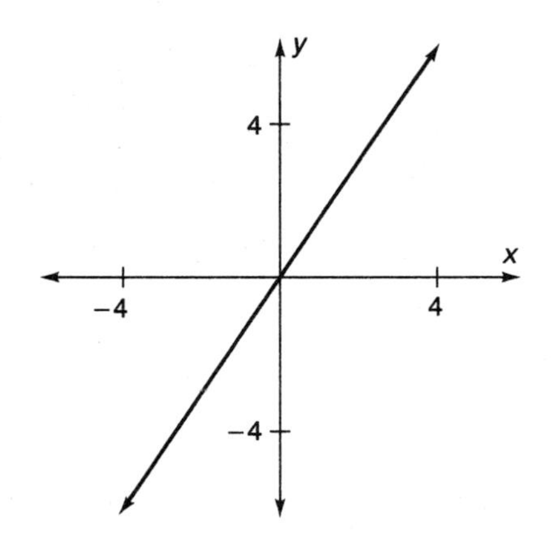

5. x-intercept: none;
y-intercept: (0, −2); slope: 0

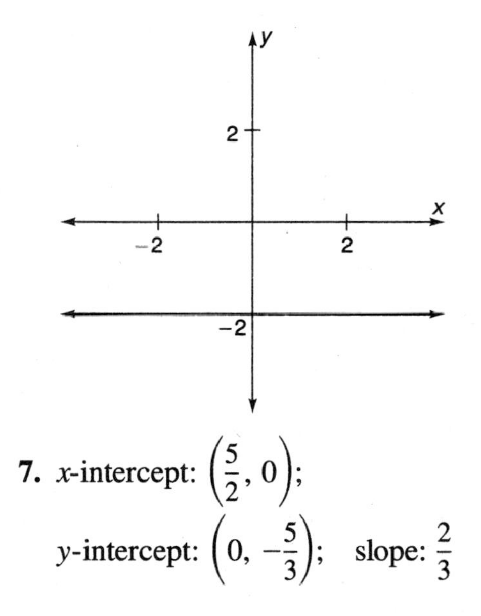

7. x-intercept: $\left(\frac{5}{2}, 0\right)$;
y-intercept: $\left(0, -\frac{5}{3}\right)$; slope: $\frac{2}{3}$

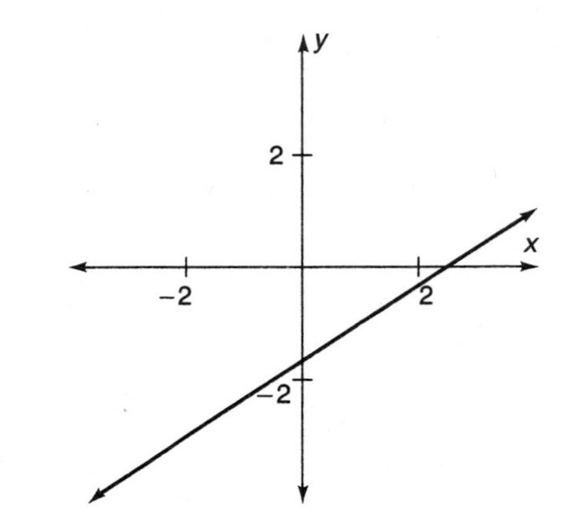

Section 2.4 (page 32)

1.

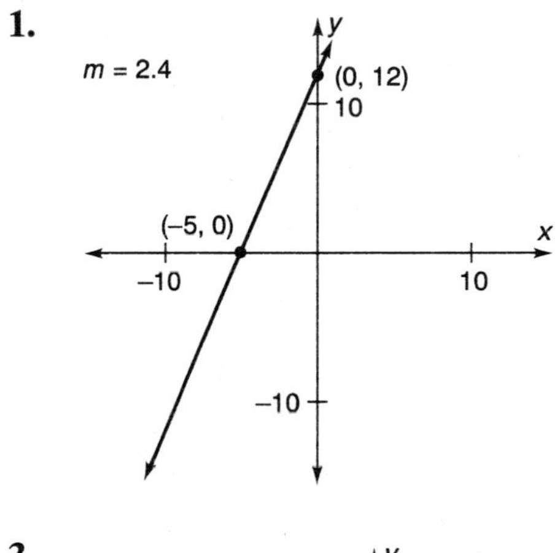

3.

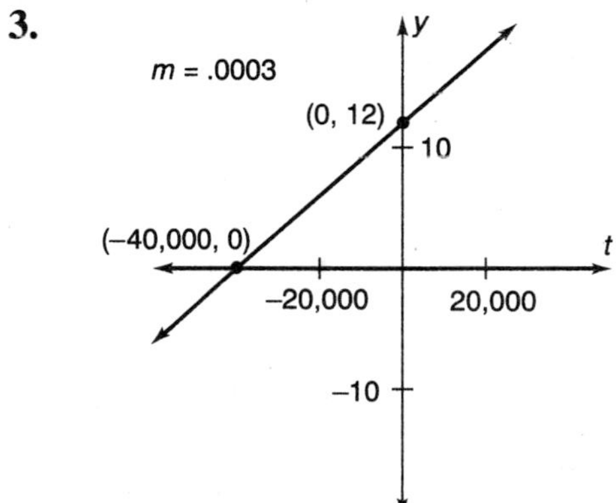

5.

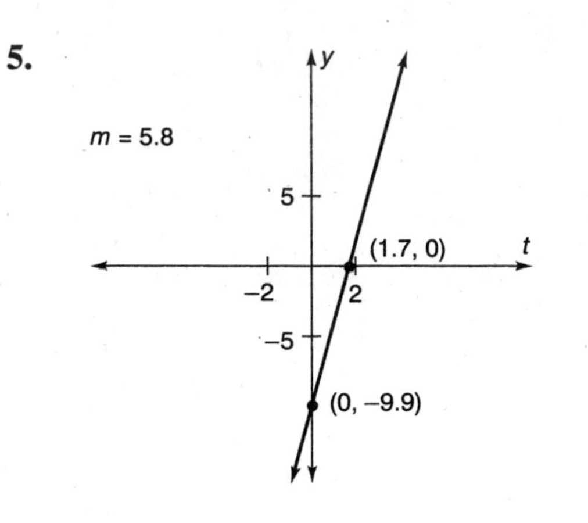

7.

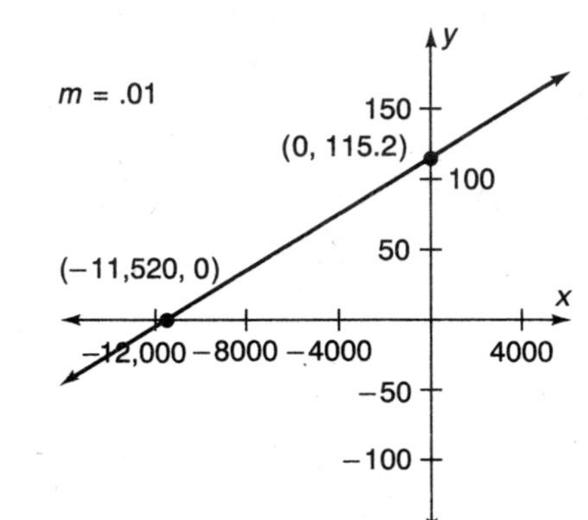

Section 2.5 (page 35)

1. $y - 5 = 4(x - 3)$
$y = 4x - 7$

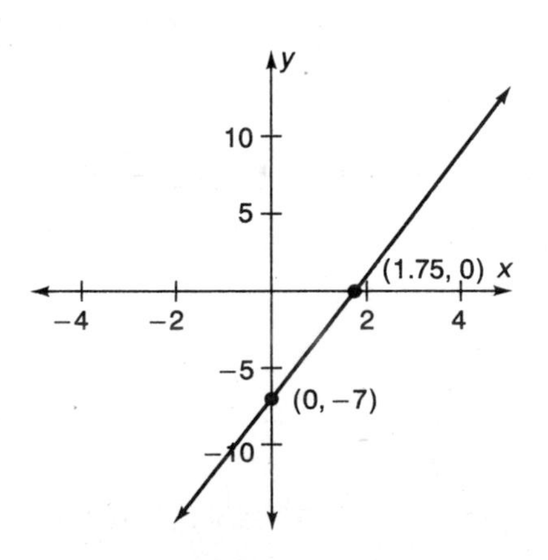

3. $y - 1 = 0(x - 3)$
$y = 1$

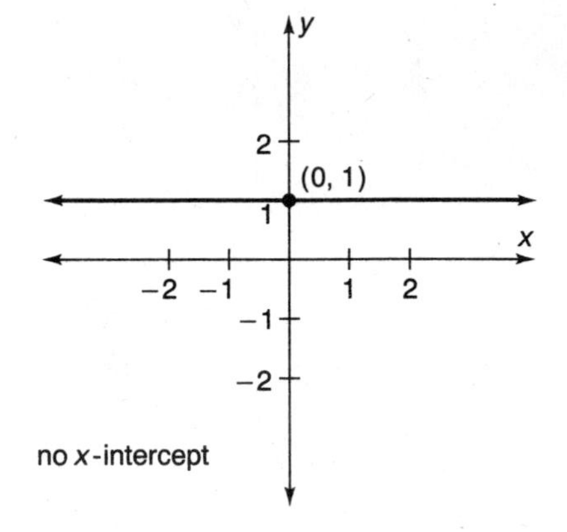

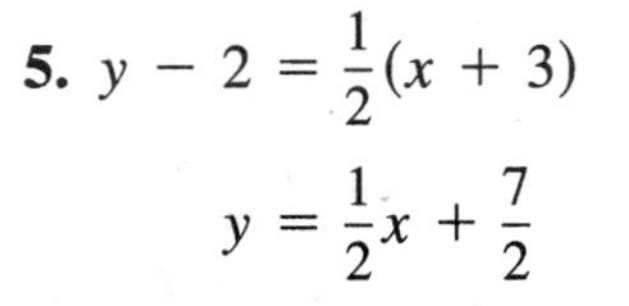

5. $y - 2 = \frac{1}{2}(x + 3)$

$y = \frac{1}{2}x + \frac{7}{2}$

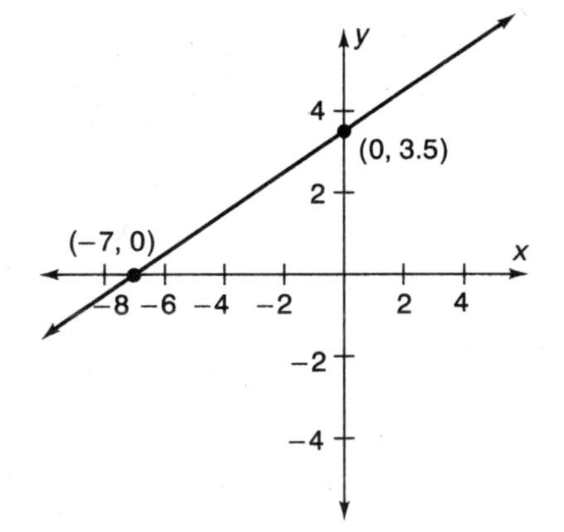

7. $y + 12 = 16.4(x)$
$y = 16.4x - 12$

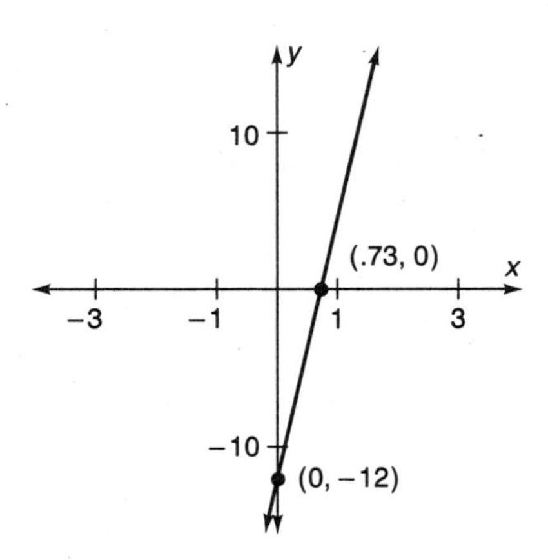

9. $y - 280 = \frac{70}{81}(x + 30)$
$y = \frac{70}{81}x + 306$

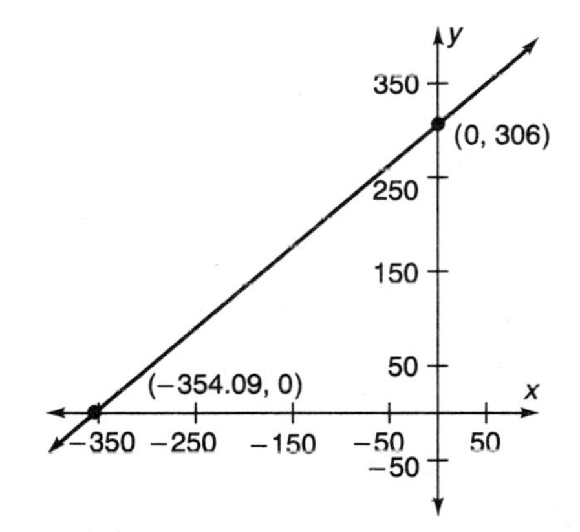

11. $y - 2.4 = 0(x - 12)$
$y = 2.4$

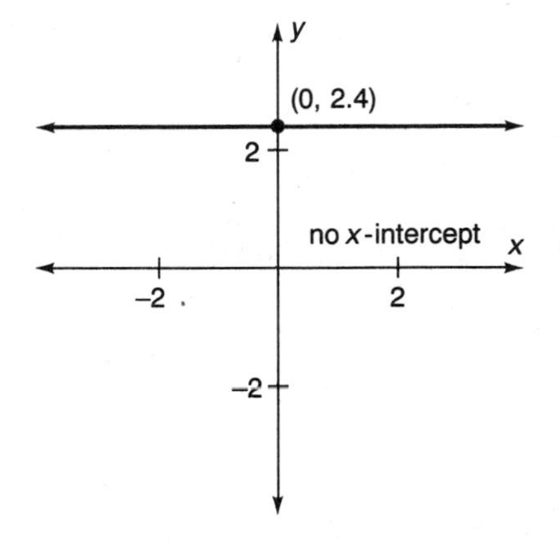

13. $y - 18.5 = .0321(x - 20)$
$y = .0321x + 17.86$

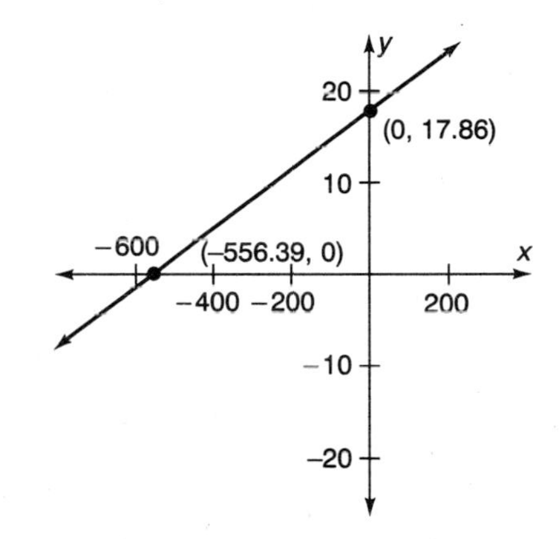

CHAPTER 4

Section 4.1 (page 53)

1. $f(g(x)) = 1 - 3x$

3. $s(r(x)) = \dfrac{5 + 2x}{x}$

5. $H(E(x)) = .13578x^3 - .00023;$
$H(E(1)) = .13555$

7. $H(A(r)) = r^3 + 5; \quad H(A(0)) = 5$

9. a) $P(F(w)) = 4 + .2w$
b) \$4.00
c) 40 gallons
d) 100 gallons

CHAPTER 6

Section 6.2 (page 77)

1. x-intercepts: (1, 0) and (5, 0); y-intercept: (0, 5). Parabola opens upward.

3. x-intercepts: (3154.5721, 0) and (.4280, 0); y-intercept: (0, 27). Parabola opens upward.

5. x-intercepts: $(-.4539, 0)$ and (9.6206, 0); y-intercept: (0, 13.1). Parabola opens downward.

7. x-intercepts: $(-1.4269, 0)$ and (1.5769, 0); y-intercept: (0, 9). Parabola opens downward.

9. x-intercepts: (0, 0) and (.2857, 0); y-intercept: (0, 0). Parabola opens downward.

11. no x-intercepts; y-intercept: (0, 3). Parabola opens upward.

Section 6.3 (page 80)

1. vertex at $(-3.1, 6.09)$

3. vertex at (.3242, 93.6306)

5. vertex at $(-.75, 5.75)$

7. vertex at (0, 0)

9. vertex at $(.125, -.0625)$

Section 6.4 (page 84)

1. min $= -1$; range $=$ all $y \geq -1$

3. min $= 110$; range $=$ all $y \geq 110$

5. min $= 0$; range $=$ all $y \geq 0$

7. max $= -14.8810$; range $=$ all $y \leq -14.8810$

Section 6.5 (page 89)

1. x-intercepts: (7.3589, 0) and $(-1.3589, 0)$; y-intercept: $(0, -10)$; vertex at $(3, -19)$ minimum at $(3, -19)$; range: all $y \geq -19$

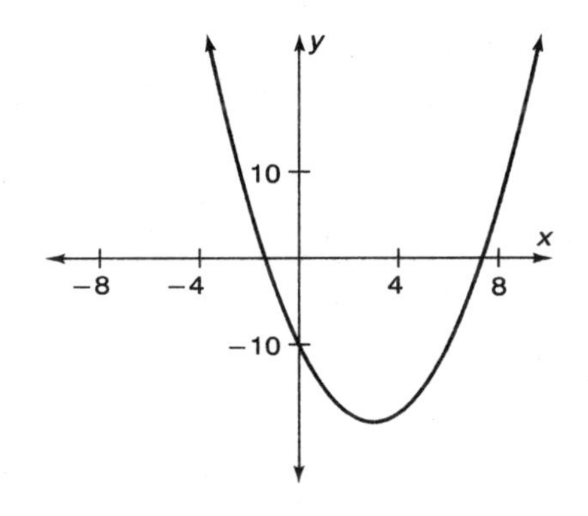

3. x-intercepts: (.8165, 0) and (−.8165, 0); y-intercept: (0, 4); vertex at (0, 4); maximum at (0, 4); range: all $y \le 4$

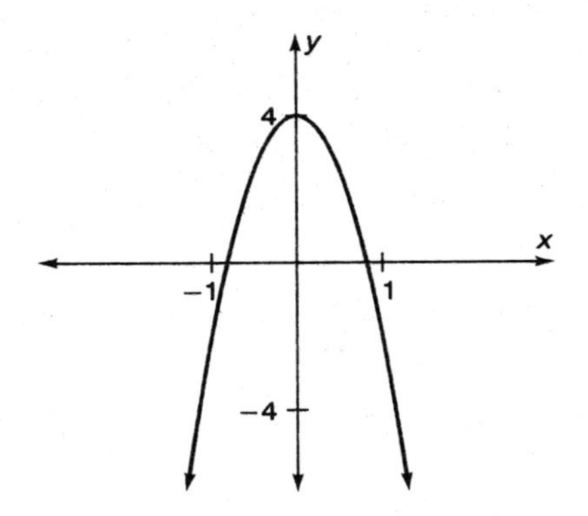

5. x-intercepts: none; y-intercept: (0, 7.3); vertex at (.20, 7.2); minimum at (.20, 7.2); range: all $y \ge 7.2$

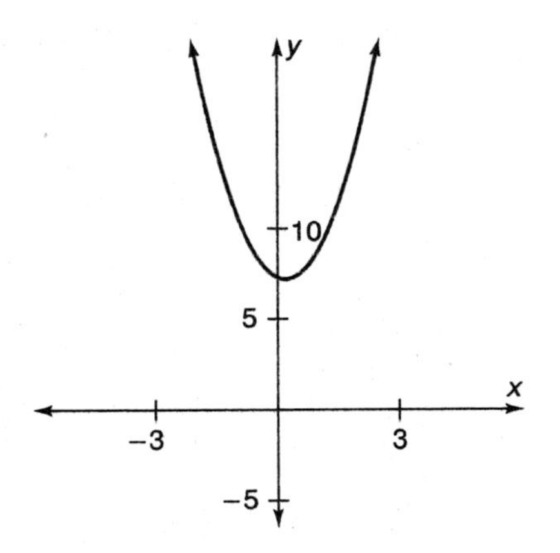

7. t-intercepts: (0, 0) and (14, 0); y-intercept: (0, 0); vertex at (7, −.49); minimum at (7, −.49); range: all $y \ge -.49$

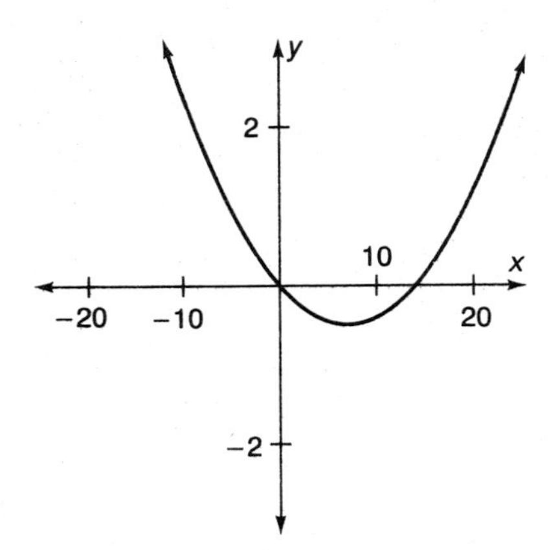

9. x-intercept: (1, 0); y-intercept: (0, −1); vertex at (1, 0); maximum at (1, 0); range: all $y \le 0$

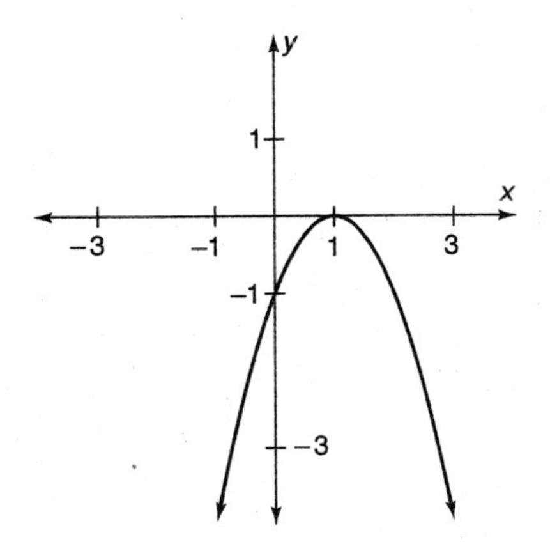

11. $x = -1.8990$ or $x = 7.8990$

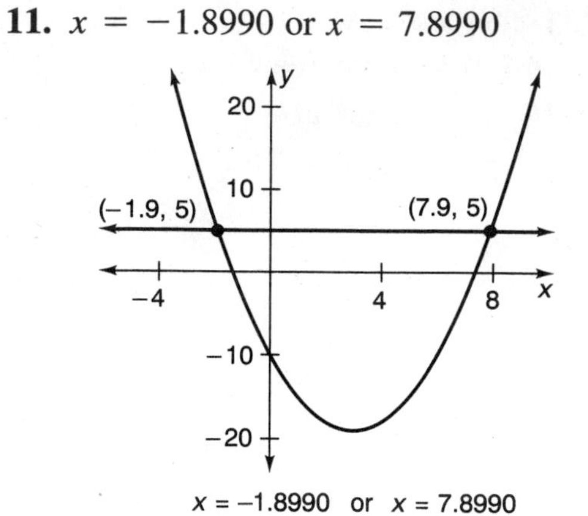

x = −1.8990 or x = 7.8990

13. No real solution.

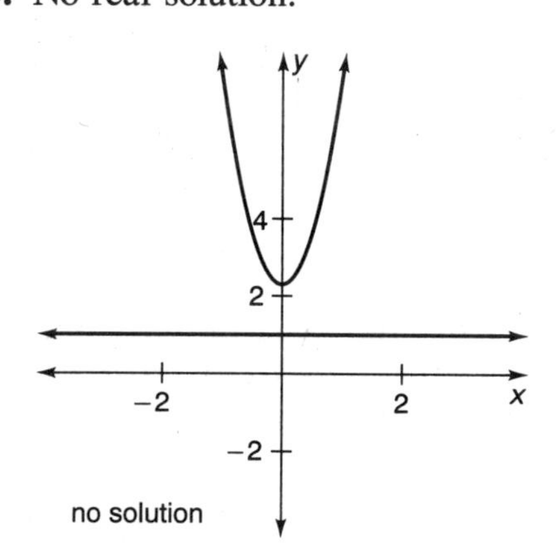

no solution

15. $x = -3.5355$ or $x = 3.5355$

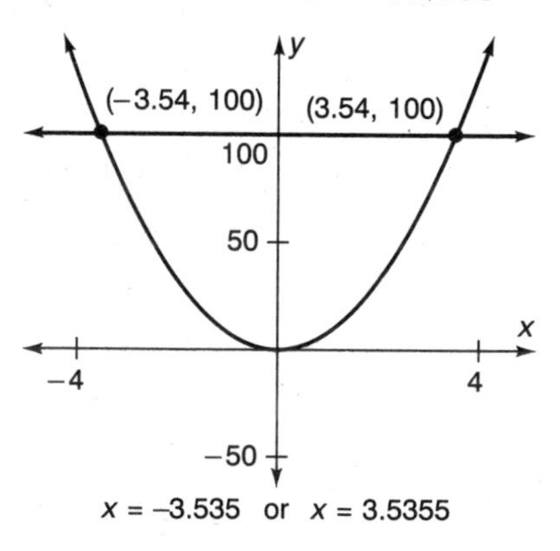

x = −3.535 or x = 3.5355

Section 6.6 (page 92)

1. vertex $\simeq (105, 6.285)$

3. vertex $\simeq (181.85, -77.81)$; $K(b) = 45$ when $x = 383.99$ or $x = -20.67$; $K(b) = 50$ when $x = 388.07$ or $x = -24.73$

CHAPTER 7

Section 7.1 (page 98)

1. $x = 1, y = 0$

3. $x = \frac{1}{5}, y = \frac{-4}{5}$

5. $x = \frac{1}{2}, y = -1$

7. $x = 2, y = 0, z = 1$

9. $x = .75, y = -1.25, z = .25$

Section 7.2 (page 103)

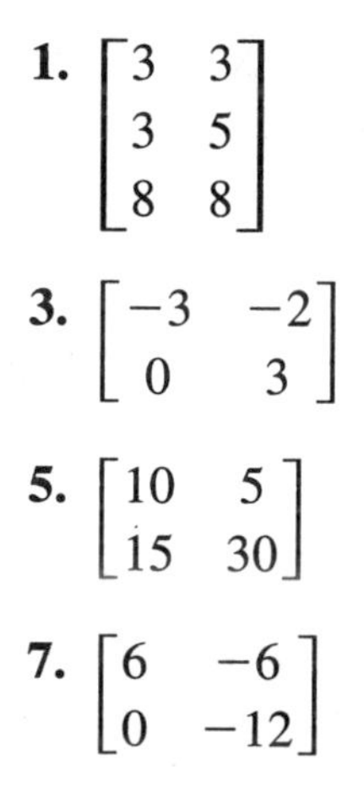

1. $\begin{bmatrix} 3 & 3 \\ 3 & 5 \\ 8 & 8 \end{bmatrix}$

3. $\begin{bmatrix} -3 & -2 \\ 0 & 3 \end{bmatrix}$

5. $\begin{bmatrix} 10 & 5 \\ 15 & 30 \end{bmatrix}$

7. $\begin{bmatrix} 6 & -6 \\ 0 & -12 \end{bmatrix}$

9. undefined

11. $\begin{bmatrix} 9 \\ 17 \end{bmatrix}$

13. undefined

15. $\begin{bmatrix} 7 \\ 7 \end{bmatrix}$

17. $\begin{bmatrix} 2 & 3 & 7 \\ 3 & 1 & 2 \\ 1 & 5 & 8 \end{bmatrix}$

19. undefined

21. not inverses, product $= \begin{bmatrix} 1 & 0 \\ 2 & 0 \end{bmatrix}$

Section 7.3 (page 110)

1. $x = 1, y = 1$

2. $x = .2, y = -.4$

5. $x = 2.2, y = .7$

7. $x = .5102, y = 4.3061$

9. $x = 2.1, y = -2.3, z = -.8$

11. $x = 4.28, y = -7.56, z = 3.2$

13. $x = .002, y = -.1, z = 14.7$

CHAPTER 10

Section 10.1 (page 144)

1. 1.8155

3. 7.3891

5. 141,667,020.5

7. y-intercept: $(0, -1)$

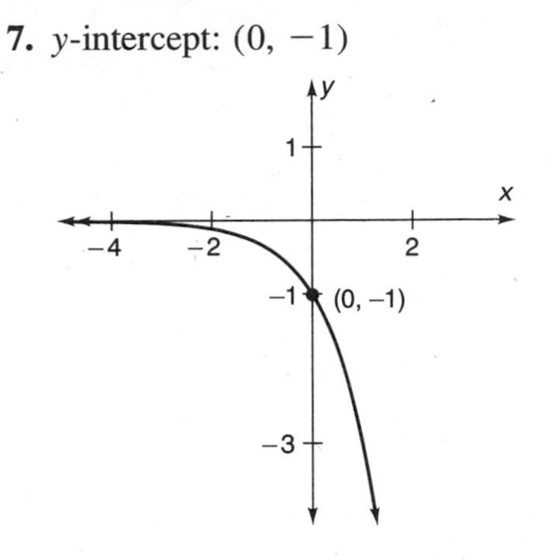

9. y-intercept: $(0, 9.6)$

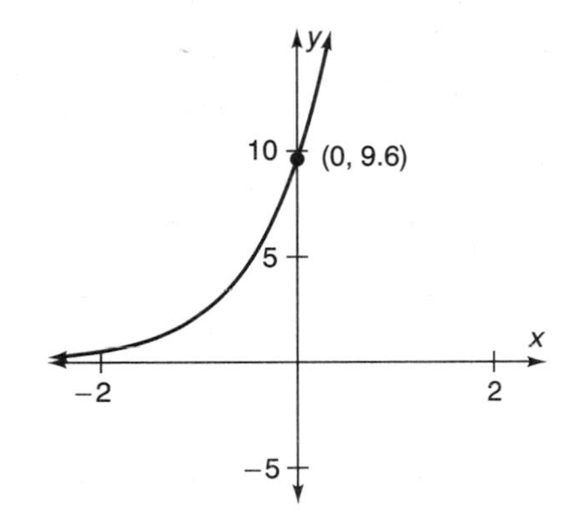

Section 10.2 (page 147)

1. $f^{-1}(y) = \dfrac{y + 1}{2}$

3. $g^{-1}(y) = 5y - 100$

5. $f^{-1}(y) = -\dfrac{y}{5.4}$

7. $x = \pm\sqrt[4]{y}$. There are two x values for every y value except $y = 0$, so the inverse does not exist.

Section 10.3 (page 153)

1. 4

3. 4

5. -1.6383

7. -3.5066

9. x-intercept: (316, 0)

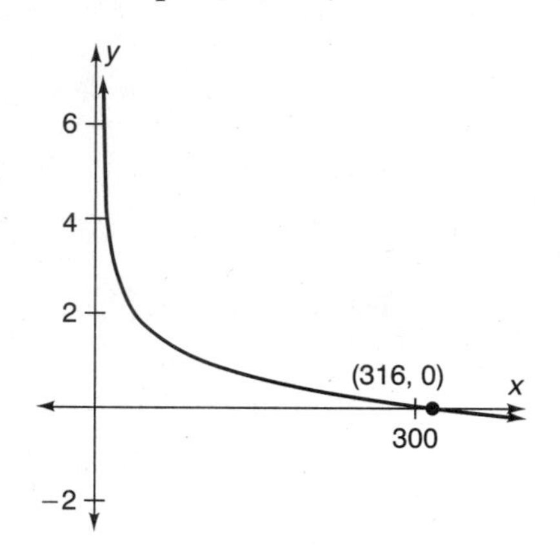

11. x-intercept: (1.3668, 0)

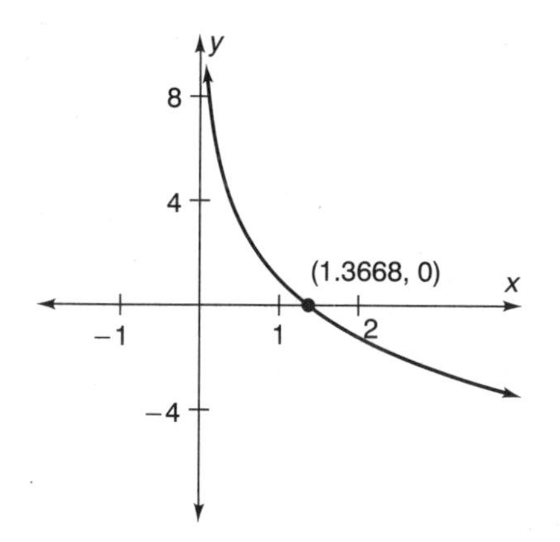

Section 10.4 (page 158)

1. $t = 8.2327$

3. $s = .3377$

5. $t = \frac{1}{2}$

Section 10.5 (page 163)

1. $g^{-1}(y) = \frac{\ln y}{\ln 3}$; $g(2) = 9$;
$g^{-1}(9) = 2$

3. $f^{-1}(y) = \frac{\ln y}{\ln 6.03}$;
$f(2.5) = 89.2880$;
$f^{-1}(2.5) = .51$

5. $f^{-1}(y) = \frac{\ln (y - 1)}{\ln 5}$; $f(0) = 2$;
$f^{-1}(2) = 0$

7. $g^{-1}(y) = e^{\frac{y - 5}{e^{-3}}}$;
$g(7) = -.8377$;
$g^{-1}(-.8377) = 7$

9. $L^{-1}(y) = 5 - e^y$;
$L(0) = 1.6094$;
$L^{-1}(1.6094) = 0$

CHAPTER 13

Section 13.1 (page 188)

1. $a = 1$; $r = 6$; $n = 7$;
$S = 335{,}923$

3. $x = 71.2$; $r = 4.3$; $n = 12$;
$S = 3{,}707{,}284{,}998$

5. $a = .5$; $r = .1$; $n = 793$;
$S = .5556$

7. $4.1 + 4.1(9.6) + 4.1(9.6)^2 + \cdots + 4.1(9.6)^7$;
$S = 34{,}391{,}828.29$

9. $.1 + .1(.01) + .1(.01)^2 + \cdots + .1(.01)^{10}$; $S = .1010$

CHAPTER 18

Section 18.1 (page 235)

1.

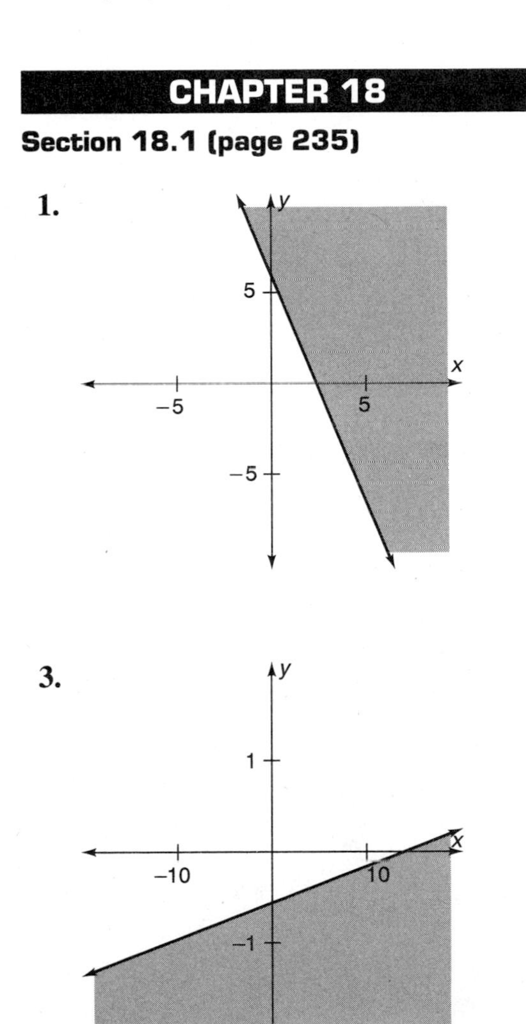

3.

5.

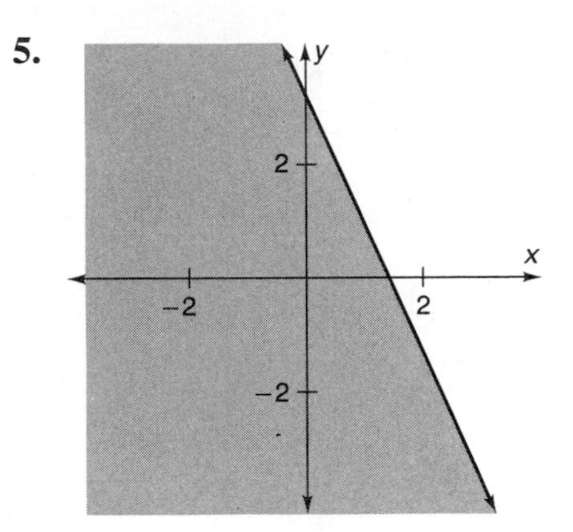

Section 18.2 (page 240)

1.

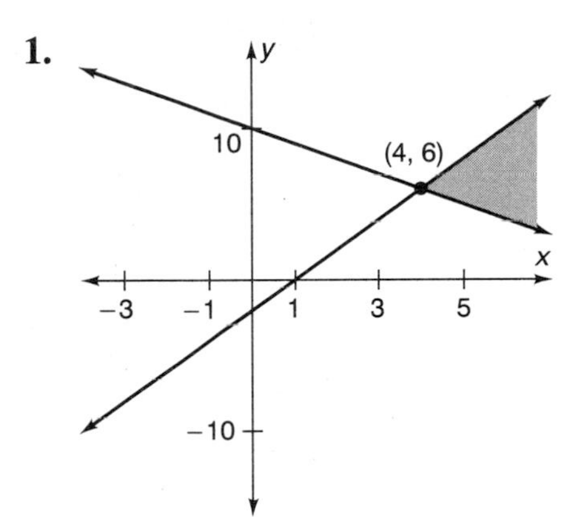

3.

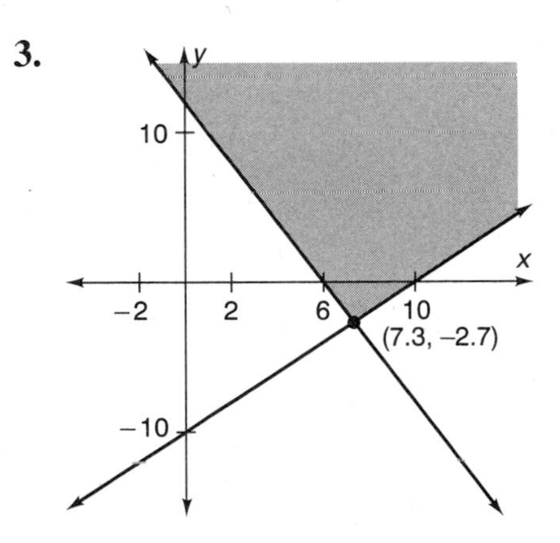

5.

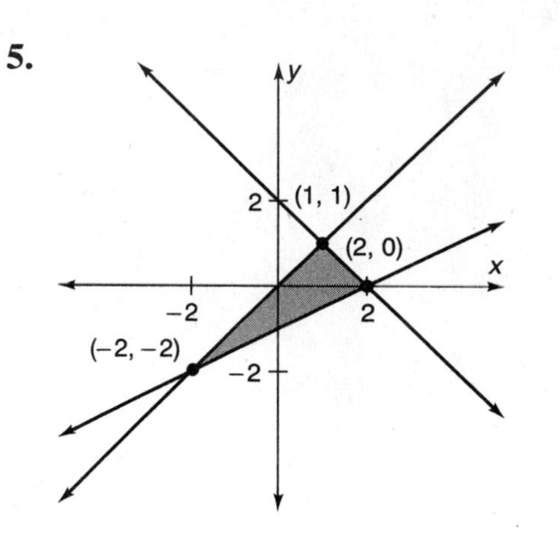

Section 18.3 (page 245)

1. $x = 3$; $y = 0$; $E = 3$

3. $x = 2$; $y = 0$; $G = 6$

5. $x = 0$; $y = 5$; $J = -5$

Acknowledgments

Unless otherwise acknowledged, all photographs are the property of Scott, Foresman and Company.

Page 1 NASA

Page 37 Kevin Schaefer/AllStock Inc.

Page 57 Tom Bean/AllStock Inc.

Page 65 Dr. Gary Settles

Page 113 Richard Pasley/Stock Boston

Page 131 Michael Collier/Stock Boston

Page 165 Matthew McVay/AllStock Inc.

Page 167 Art Wolfe/AllStock Inc.

Page 175 Bob Daemmrich/Stock Boston

Page 181 Nick Gunderson/AllStock Inc.

Page 195 Tim David/AllStock Inc.

Page 201 Jim Pickerell/Gamma-Liaison

Page 209 Ellis Herwig/Stock Boston

Page 213 Michael Dwyer/Stock Boston

Page 227 James Randklev/AllStock Inc.

Page 247 Milt & Joan Mann/Cameramann International, Ltd.

Page 257 Bob Daemmrich/Stock Boston

Bibliography

The references listed here are not only source books used for information in this text, but also include some additional related readings which should be of interest.

Abbey, Edward. *Desert Solitaire* (New York: Simon & Schuster, 1968).

Bates, Albert. *Climate in Crisis: The Greenhouse Effect and What We Can Do* (Summerton, Tenn.: The Book Publishing Company, 1990).

Berry, Wendell. *What are People for?* (Berkeley, California: North Point, 1990).

Brown, Lester. *State of the World 1991: Worldwatch Institute Report on Progress Toward a Sustainable Society* (New York: Norton, 1991).

Council on Environmental Quality, *Environmental Quality,* (Washington DC: US Govt. Printing Office, 1992).

Council on Environmental Quality, *Environmental Trends,* (Washington DC: US Govt. Printing Office, 1989).

Ehrlich, Gretel. *The Solace of Open Spaces* (New York: Viking, 1985).

Falk, Jim and Andrew Brownlow. *The Greenhouse Challenge: What's to Be Done?* (Ringwood, Victoria: Penguin Books Australia, 1989).

Foreman, Dave. *Confessions of a Eco-Warrior* (New York: Harmony, 1991).

Gaia. *An Atlas of Planet Management* (New York: Anchor Press/Doubleday, 1984).

Global Tomorrow Coalition. *The Global Ecology Handbook* (Boston: Beacon Press, 1990).

Gribben, John. *Hothouse Earth: The Greenhouse Effect and GAIA* (New York: Grove Weidenfeld, 1990).

Harte, John. *Consider a Spherical Cow: A Course in Environmental Problem Solving* (Mill Valley, CA: University Science Books, 1988).

Karlesky, Joseph. *Thinking About Environmental Policy,* HarperCollins, (New York, 1992).

Kaufman, Donald, and Franz, Cecelia. *Biosphere 2000: Protecting our Global Environment,* HarperCollins College Publishers, (New York, 1994).

Lamb, Marjorie. *Two Minutes a Day for a Greener Planet* (New York: HarperCollins, 1991).

Leggett, Jeremy (editor). *Global Warming, the Greenpeace Report* (Oxford: Oxford University Press, 1990).

Levy, Walter and Hallowell, Christopher. *Green Perspectives: Thinking and Writing About Nature and the Environment,* HarperCollins College Publishers, (New York, 1994).

Lopez, Barry. *The Rediscovery of North America* (Lexington, Ky., University of Kentucky, 1990).

Lorde, Audre. *The Marvelous Arithmetics of Distance,* W. W. Norton & Company, (New York, London, 1993).

Merwin, W. S. *The Rain in the Trees* (New York: Alfred A. Knopf, 1988).

Mitchell, Finis. *Wind River Trails* (Salt Lake City: Wasatch, 1975).

Nabokov, Peter (editor). *Native America Testimony* (New York: Viking, 1985).

Oliver, Mary. *New and Collected Poems* (Boston: Beacon Press, 1992).

Rich, Adrienne, *An Atlas of the Difficult World* (New York: Norton, 1991).

Roan, Sharon. *Ozone Crisis* (New York: John Wiley, 1989).

The Earth Works Group. *50 Simple Things You Can do to Save the Earth* (Berkeley, Calif., 1989).

The Student Environmental Action Coalition. *The Student Environmental Action Guide: 25 Simple Things We Can Do* (New York: HarperCollins, 1991).

Tietenberg, Tom. *Environmental and Natural Resource Economics,* 3rd edition (New York: HarperCollins, 1992).

U.S. Bureau of the Census, *Statistical Abstract of the United States* (Washington, D.C.: U.S. Government Printing Office, 1993).

U.S. Bureau of the Census, *Historical Statistics of the United States, Colonial Times to 1970* (Washington, D.C.: U.S. Government Printing Office, 1976).

U.S. Congress, Office of Technology Assessment, *Changing by Degrees: Steps to Reduce Greenhouse Gases* (Washington, D.C.: U.S. Government Printing Office, 1986).

U.S. Department of Energy, Energy Information Administration, *The Motor Gasoline Industry: Past, Present, and Future* (Washington, D.C.: U.S. Government Printing Office, 1991).

Worldwatch Institute, *Worldwatch Papers* (Washington, D.C.).

Worldwatch Institute, *State of the World, 1993, State of the World, 1994,* W. W. Norton & Company, (New York, London, 1993, 1994).

World Resources Institute, *World Resources 1990–91* (New York, Oxford: Oxford University Press, 1990).

World Resources Institute, *World Resources 1992–93,* (New York, Oxford: Oxford University Press, 1992).

World Resources Institute, *The 1992 Information Please © Environmental Atlas* (Boston: Houghton Mifflin Company, 1992).

World Resources Institute, *The 1994 Information Please © Environmental Atlas* (Boston: Houghton Mifflin Company, 1994).

Index